Rewilding Food and the Self

"This is an original and thought-provoking book about practices and meanings of 'rewilding' food across Europe. Cases of hunting game, foraging in the countryside, dumpster-diving in the city and producing 'natural' wine uncover how fascination with the wild highlights key moral, environmental, social and health issues in contemporary eating practices."

Carole Counihan, *Editor-in-Chief, Food and Foodways*

"There is, in the attraction for the 'wild,' a will to break with the established order. This theme regularly reappears in current social life and has evolved according to historical contexts. '*Rewilding Food and the Self: Critical Conversations From Europe*' focuses on the contemporary version and analyzes its current and timeless social issues."

Jean-Pierre Poulain, *Université de Toulouse, France*

This volume contributes to the return to nature movement that is very much in vogue in contemporary European societies, by examining the place of food and eating in the "rewilding" process.

It is divided into three parts, each of which consists of conversations between social scientists, with fieldwork collected from across Denmark, Finland, France, Italy, Norway and Switzerland. The first part focuses on the ways in which the hunter-gatherer livelihood has been transformed into a resilient, simpler and ecological way of life. It is dedicated to hunting and identifies the contexts in which large wild game meat is consumed and the reasons why such a product is still valued today. The second part shows how some practices that aim to reconnect with natural processes are developing within a market economy. Case studies on natural wine and fasting retreats help us to identify the promises that producers and promoters are relying on in order to disseminate them. Finally, the third part considers how this process of rewilding food is expressed in post-modernity. By focusing on two normative frameworks in which the rhetoric of the wild is mobilized although it is not expected to be in these terms – urbanity and the gender order – the goal is to understand the extent to which referring to the wild in food discourses and practices contributes to challenging our identities, and to creating possible forms of emancipation.

This book will be of great interest to students and scholars interested in food cultures, human–nature relationships and sustainable diets.

Tristan Fournier is a sociologist and research fellow at the French National Centre for Scientific Research (CNRS – Iris, Paris).

Sébastien Dalgalarrondo is a sociologist and research fellow at the French National Centre for Scientific Research (CNRS – Iris, Paris).

Routledge Studies in Food, Society and the Environment

True Cost Accounting for Food
Balancing the Scale
Edited by Barbara Gemmill-Herren, Lauren E. Baker and Paula A. Daniels

Holiday Hunger in the UK
Local Responses to Childhood Food Insecurity
Michael A. Long, Margaret Anne Defeyter and Paul B. Stretesky

School Farms
Feeding and Educating Children
Edited by Alshimaa Aboelmakarem Farag, Samaa Badawi, Gurpinder Lalli and Maya Kamareddine

The Vegan Evolution
Transforming Diets and Agriculture
Gregory F. Tague

Food Loss and Waste Policy
From Theory to Practice
Edited by Simone Busetti and Noemi Pace

Rewilding Food and the Self
Critical Conversations From Europe
Edited by Tristan Fournier and Sébastien Dalgalarrondo

For more information about this series, please visit: www.routledge.com/books/series/RSFSE

Rewilding Food and the Self

Critical Conversations From Europe

Edited by Tristan Fournier and Sébastien Dalgalarrondo

Routledge
Taylor & Francis Group
LONDON AND NEW YORK

earthscan
from Routledge

First published 2023
by Routledge
4 Park Square, Milton Park, Abingdon, Oxon OX14 4RN

and by Routledge
605 Third Avenue, New York, NY 10158

Routledge is an imprint of the Taylor & Francis Group, an informa business

British Library Cataloguing-in-Publication Data
A catalogue record for this book is available from the British Library

Library of Congress Cataloging-in-Publication Data
Names: Fournier, Tristan, editor. | Dalgalarrondo, Sébastien, editor.
Title: Rewilding food and the self : critical conversations from Europe / edited by Tristan Fournier and Sébastien Dalgalarrondo.
Description: New York : Routledge, [2023] | Includes bibliographical references and index.
Identifiers: LCCN 2022020424 (print) | LCCN 2022020425 (ebook) | ISBN 9781032152912 (Hardback) | ISBN 9781032152936 (Paperback) | ISBN 9781003243496 (eBook)
Subjects: LCSH: Food—Europe. | Wild foods—Europe. | Restoration ecology—Europe.
Classification: LCC GT2853.E8 R496 2023 (print) | LCC GT2853.E8 (ebook) | DDC 394.1/2094—dc23/eng/20220816
LC record available at https://lccn.loc.gov/2022020424
LC ebook record available at https://lccn.loc.gov/2022020425

ISBN: 978-1-032-15291-2 (hbk)
ISBN: 978-1-032-15293-6 (pbk)
ISBN: 978-1-003-24349-6 (ebk)

DOI: 10.4324/9781003243496

Typeset in Goudy
by Apex CoVantage, LLC

Contents

Contributors

Ophélie Bidet is a PhD candidate in sociology at the University of Neuchâtel (Switzerland). Her dissertation focuses on the evolution of the practice of fasting outside the religious context and its inclusion in lifestyles.

Sébastien Dalgalarrondo is a research fellow in sociology at the CNRS (the French National Centre for Scientific Research) and a member of the Institut de Recherche Interdisciplinaire sur les enjeux Sociaux (Iris) in Paris (France). His first work focused on the role of patient associations in the pharmaceutical innovation process. The second part of his research is based on the analysis of professional sports, anti-ageing medicine and "rewilding" practices in order to elaborate a critique of the "optimal man" contemporary promise.

Sergio Dalla Bernardina is a professor of ethnology at the Université de Bretagne Occidentale (France). He is a member of the IIAC and responsible for two seminars at the École des Hautes Études en Sciences Sociales (EHESS), Paris. His research focuses on the relationship between Man and the environment, the animal question and the conditions of production of anthropological discourse. He is the author of several books. He runs the blog: http://lanimal commepretexte.blogspot.fr.

Nicoletta Diasio is an anthropologist and sociologist, a full professor at the University of Strasbourg (France) and a member of the Laboratoire interdisciplinaire en études culturelles (LinCS 7069) and of the Institut universitaire de France. Her research focuses on body, age transitions, gender, food and chronic illness.

Sandrine Dury is an agricultural economist specialized in food consumption at the Cirad (Montpellier, France). She is studying food systems in interdisciplinary designs. She is working on the linkages between food security of rural households, farm production and access to markets and to natural resources in different places of the world, mainly in African countries.

Vulca Fidolini is an associate professor of sociology at the University of Lorraine (France). He is also a member of the Laboratoire Lorrain de Sciences Sociales (2L2S). His main current research topics are gender (masculinities), age transitions, health and food.

Tristan Fournier is a research fellow in sociology at the CNRS (the French National Centre for Scientific Research) and a member of the Institut de Recherche Interdisciplinaire sur les enjeux Sociaux (Iris) in Paris (France). His work focuses mainly on the social, moral and political issues of food changes, and questions the status of biomedical knowledge and the relationships between food, health and the body.

Olivier Lepiller is a sociologist working at the Cirad, MoISA (Montpellier, France). His research focuses on the social change of food norms and practices, the food transitions toward more sustainable food systems, and the roles of social innovations, critics of industrial food and local policies in these transitions.

Léo Mariani is an anthropologist, a teacher-researcher with the Muséum National d'Histoire Naturelle and a member of the Unité Mixte de Recherche Eco-anthropologie with the Musée de l'Homme in Paris (France). He is interested in researching how certain plants become food, in France and Asia, using an ontogenetic approach to examine the production of worlds in space and time. In his work, he brings together the anthropology of knowledge and the anthropology of nature.

Sabine Mégarbané is a graduate of the École des Hautes Études en Sciences Sociales (EHESS, Paris, France) where she completed a master's degree in sociology. She is interested in environmental movements, wastes and practices of resistance to the capitalist system. In 2021, she completed a dissertation on informal waste recovery practices by freegans in France.

Flaminia Paddeu is an assistant professor in geography at the Sorbonne University Paris Nord and at Pléïade laboratory (France). She analyzes the socio-ecological issues of urban agriculture, gleaning and foraging in northern metropoles (France and United States).

Christelle Pineau holds a PhD degree in social anthropology and ethnology from the École des Hautes Études en Sciences Sociales (EHESS, Paris, France). She is a research associate at IIAC and IRM (University of Bordeaux). Her work focuses on the field of wine, agroecologies, the articulation of sensitive and academic knowledge, and power relations.

Gun Roos is a senior researcher at Consumption Research Norway (SIFO) at the Oslo Metropolitan University (Norway). She holds a PhD degree in nutritional anthropology from the University of Kentucky and an MS degree in nutrition from the University of Helsinki. Her research interests include social and cultural aspects of food, inequalities, gender and policy.

Ophélie Roudelle is a project manager at UNESCO Chair in World Food Systems (Cirad, Montpellier, France). She is in charge of disseminating, communicating and promoting the science-society dialogue on the Urbal research project, which proposes a participatory method for evaluating the impacts of food innovations.

Fabien Roussel is an assistant professor of geography at the University of Artois (France) and at Discontinuités Laboratory. He studies the socio-ecological issues raised by plants and flora in northern metropolises (France and the Netherlands).

Acknowledgments

We would like to thank the *LabEx Tepsis* (https://tepsis.io/) and our research laboratory, the *Institut de Recherche Interdisciplinaire sur les enjeux Sociaux* (Iris, Paris), for their financial support for this book.

We also thank *Cadenza Academic Translations* for the translation of several chapters.

Introduction

Food, self and search for the wild[1]

Tristan Fournier and Sébastien Dalgalarrondo

From local to wild: the hypothesis of a new food narrative

"The future belongs to nostalgia"[2]: this is one of the lessons that Claude Fischler and Estelle Masson (2008) drew from their comparative study on practices and representations relating to food in five European countries (England, Germany, France, Italy and Switzerland) and the United States. Of the 6,000 people in their sample, 56% thought that food products were less healthy than 40 or 50 years ago. More generally, 62% evoked a degradation of food habits. From the sensory point of view, the French seemed most nostalgic: 61% of them thought that food did not taste as good as it used to. Of course, these elements covered, sometimes, significant social and cultural differences, but the survey revealed a strong trend: "nostalgia is dominant." In the collective work based on this broad survey, other European researchers provided some specific information about their countries. Alan Beardsworth evoked a "food competence nostalgia" in the United Kingdom. Here, the term "competence" refers to a capacity acquired through diffuse impregnation, which formerly offered the advantage of requiring no conscientization about food and eating, but which had become eroded as nutritional knowledge developed. Hence, the reference to a "feeling of lost innocence" and the emergence of a form of "nostalgia for this innocence, thanks to which a competent eater could experience choice, preparation, and consumption as instinctive, rather than problematic" (Beardsworth 2008). Eva Barlösius showed that, in Germany, one of the models making up the food morality that she studied was linked to an idealization of the past: some of the eaters interviewed spontaneously saw and presented food in the preindustrial period as healthier, more natural and tastier than food today. Moreover, this fantasy past referred to the sharing of common values, to more collective modes of food consumption and regulation. It is this generalized loss of meaning that led some of her sample to idealize the past and (in her words) to "glorify" and "romanticize" traditions (Barlösius 2008).

1 Translated by Cadenza Academic Translations.
2 Translator's note: Unless otherwise stated, all translations of cited foreign-language material in this chapter are our own.

DOI: 10.4324/9781003243496-1

When it comes to food, it seems that views in contemporary European societies are partly characterized by the maxim "it was better before." This romance with the past, which Zygmunt Bauman (2017) conceptualized as "retrotopia," is not new, however. Jean-Pierre Poulain (2017), drawing on the works of Edgar Morin (1975) and particularly on his concept of the "neo-archaic mentality," showed that in response to food products and practices starting to become more standardized, the late 1960s saw emerging enthusiasm for local and authentic food, with the establishment of a "rustic gastronomy" in France. In a way, this gastronomy had broken free from the principles of gourmet cuisine, seen as too complex and distinctive (Bourdieu 1984). Instead, it embraced "natural values," which focused on the simplicity, naturalness and origin of the products, and which would soon become defenses against the rise of processed food. Concerning the current situation, Poulain states that "contemporary interest in local regional cuisines is part of the nostalgia for a 'social space' where one might eat untroubled by anxiety, sheltered by a clearly identifiable, and identifying, culinary culture" (Poulain 2017, 13). In fact, eating local is seen not only as a way of promoting products grown or reared nearby but also, and above all, as a way of getting to know the people who produce our food and therefore of reintroducing proximity (and with it trust and transparency) into the connections between producers and consumers. This trust helps to reduce social complexity and particularly the blind spots of the market (Luhmann 1979). Moreover, it is seen as a way of reconnecting with the values attributed to rurality, the popularity of which seems to be growing as societies become more urbanized. This is a case of the local coming to the rescue of nostalgia. Since the 1960s, this quest for meaning, health and authenticity – let's say for "tradition" (Sébastia 2017) – has progressively become entwined with a denunciation of the harmful effects of the agro-industrial food system and has therefore become politicized (Counihan and Siniscalchi 2013). It is now embodied in initiatives on various levels, from the local level through alternative food networks such as AMAPs[3] (Lamine 2005) to the transnational level with social movements like Slow Food (Siniscalchi 2020).

In their conclusion, Fischler and Masson (2008) encouraged us to "re-enchant food," that is, to put the magic back into the food we eat. Conceivably, this great narrative of the local and authentic was able to play this role for a time. But what if it is no longer enough today? Since their study around 15 years ago, a surprise guest has come to the table, at least for more reflexive eaters: the environmental crisis. What was previously just one of many factors in dietary choices has progressively become an urgent consideration, a requirement that could increase anxiety and intensify reflexivity: the fear of choosing the wrong food, sorting waste

3 AMAP: Association pour le Maintien de l'Agriculture Paysanne (Association for the Preservation of Small-Scale Agriculture). AMAPs are a form of community-supported agriculture that aims to protect small-scale agriculture near towns, by creating a short supply chain between producers and consumers, in which consumers pay in advance for a weekly food basket from the producer. ("Association pour le maintien d'une agriculture paysanne," Pour la solidarité, accessed March 11, 2022, www.ess-europe.eu/fr/bonnepratique/association-pour-le-maintien-dune-agriculture-paysanne).

incorrectly and falling victim to greenwashing. This is where doubt takes hold. What is the concrete reality behind the products we buy at the market or from "small producers," the quality of which seems to depend primarily on physical or moral attachments? Are they really good for us and for the environment? Can we trust them? Perhaps the reassuring model of "rustic gastronomy" is natural only by virtue of the value we give to it.[4] In view of the climate and environmental crisis, cracks may be appearing in the rural model of the preindustrial period evoked by Barlösius (2008), which may soon no longer be a sufficiently distant and inspiring horizon to satisfy our nostalgia and our need for authenticity and naturalness.

Therefore, the hypothesis on which we want to work in this book is that of the emergence of a new food narrative built around the theme of the wild. Practices that are currently still marginal and distinctive are spreading. Recently, you have probably heard about the "paleo diet"[5] or about Noma (Copenhagen, Denmark) being named the "world's best restaurant" for the nth time, for its menus based around mushrooms and game. Or perhaps, at the supermarket, you have come across products whose main selling point is their ancestral use by "primitive" food cultures. Or online you may have seen courses on learning about and cooking with edible wild plants. So what do these products and practices have in common? Some would evoke the worrying commodification of nature and the essentialization of the moral qualities attributed to it, and justifiably so. However, it is the wild dimension of these products and practices that we want to examine more specifically here. In French, the term for *wild* is *sauvage*, referring to the Latin *sylvae*, which means *forest*. The wildness of a living being seems therefore to be etymologically connected to its physical or symbolic proximity with the world of the forest. Thus, the wild refers to an ontological alterity that kindles fantasies and utopias (Dalla Bernardina 2012; Dalgalarrondo and Fournier 2020), by representing that which is pure, original, natural and even good. This is where the connection with food comes into play. For a moment, let us return to the study by Fischler and Masson (2008): the metaphor most cited (47%) as the best expression of the relationship that the 6,000 people in the sample said they had with their food was that of . . . the tree![6] In the survey participants' minds, the tree referred not only to nature, of course, but also to local identity and to what we all have fundamentally in common, that is, to an anchoring or (more precisely) a rootedness in a territory, a region or a culture. The tree symbolized the balance between eaters and their natural, social and cultural environment. However, it seems that the environmental crisis is disrupting this imaginary: the tree, which is part of the landscape in both the literal and figurative senses, is threatened. After all, is it not in danger of disappearing, of being uprooted? The practices evoked earlier could therefore be part of a movement of rewilding the

4 Olivier Lepiller (2012) showed very clearly that the value of naturalness accorded to a food product is always fragile, debatable and changing and that its distinction from artifice is never self-evident.

5 The aim of the "paleo diet" is to get as close as possible to the diet of Paleolithic hunter-gatherers. This means choosing animal products (meat, fish and eggs), vegetables, fresh fruits and nuts, and avoiding food from agriculture and industry, such as cereals, legumes and dairy products.

6 This was followed by the temple (20%), the car (17%) and the factory (16%).

self and food, which (even more than the narrative of the local) seems to promote the search for untouched purity and to be based on even greater rootedness, that is, on "enforesting" ourselves, sinking our roots ever deeper into history and time. Thus, the pre-Neolithic period becomes a powerful source of inspiration, and our hunter-gatherer ancestors are seen as role models, equipped with instincts, treading lightly on the Earth, and feeding parsimoniously on its fruits (and its animals). The imaginary of the wild, and all its accompanying contradictions, could embody this new food narrative that would make it possible to simultaneously embrace a nostalgia, a quest for meaning, naturalness and purity, the search for an *in-common*, and growing attention to nutrition and the environment.

Importing the "rewilding" concept into the world of food

This book is a follow-up to a series of conferences[7] that we organized, during which we tried to understand the manifestations of the wild and their meanings in contemporary eating. More broadly, we were also trying to understand what this trend said about us as eaters, citizens and individuals. We thought that the term *rewilding* described this movement well. This term was first conceptualized in the field of ecology, where it is closely related to the issues of biodiversity and the founding experiments with reintroducing wild animals (e.g., in Yellowstone, USA, in 1995). In the last 20 years, this concept has become a new paradigm of nature conservation, built around the leitmotif of "let nature take its own course." The idea behind this is that nature is "well made" and inherently beneficial, and that it has a neutral and stable point of origin toward which we should be aiming. We can identify several ways in which the domain of food borrows from this concept, which have helped to draw attention to the importance of reconsidering nature's ecological processes when it comes to agriculture (Carolan 2015; Shiva 2020). In this book, we want to extend this thinking to dietary and food-related practices themselves and to the meaning that we give to this rewilding movement. In doing so, we want to consider the possibilities, stakes and meaning of such a transposition of the concept in a more concrete way. We believe that three key threads can be identified within this issue.

Rewilding one's body

The way in which environmental movements mobilize the concept of rewilding testifies to a need to break away from an anthropocentric understanding of

7 The first conference held in March 2019 in France at the École des Hautes Études en Sciences Sociales (EHESS) (School of Advanced Studies in the Social Sciences) in Paris. The title of the conference, which took place in French, was "Sauvages! Promesses et pratiques contemporaines de l'ensauvagement" ("Wild! Contemporary Promises and Practices of Rewilding"). We then organized a panel titled "Rewilding Food and the Self" within the Association for the Study of Food and Society (ASFS) and the Agriculture, Food and Human Values Society (AFHVS) international conference held at the University of Alaska Anchorage (USA) in June 2019.

nature (Descola 2013) that regarded humans as external to it, and logically its masters, owners and controllers. This shift of perspective (from anthropocentrism to biocentrism) is inspiring for us: it is apparently no longer just about rewilding nature but also about rewilding the human body through food. Thus, a principle of preserving territories and species is replaced by that of optimizing a body and physiological potential, getting as close as possible to nature and to *our* nature. Dietary reflexivity is combined here with a concern for health. This dynamic has been analyzed, in a way, from the angle of "nutritional primitivism" (Knight 2015; Loyer and Knight 2018): a dietary reflexivity based on the purity of the wild and an aversion to the processed, and which is characterized by a critique of the agro-food industry and a strong valorization of the pre-Neolithic way of life. Thus, new ways of seeing the human body emerge. These are based on mismatch theory: according to many researchers who align themselves with evolutionist thought, the physiological adaptations acquired during the long period when *Homo sapiens* was a hunter-gatherer are serious handicaps today, in our societies where food is abundant. The Neolithic Revolution, which brought about settlement (the rise of agriculture and livestock farming), is therefore often seen as the turning point of human evolution. So-called diseases of civilization, such as obesity, type 2 diabetes and certain cancers, are seen as the modern illustration of this: our physiology, long shaped by an environment in which people experienced periods of shortage, fluctuation and food insecurity (Paleolithic hunter-gatherers probably did not eat three meals a day at set times![8]), is seen as ill-adapted to modern life. In fact, our bodies are still better suited to dealing with shortage than to handling abundance. It is this evolutionary trap that has led certain authors to suggest that diseases of civilization (particularly obesity) could be the fate of a few sacrificed generations, until biological and physiological mechanisms adjust to this new food environment (Poulain 2009).

On the other hand, some authors have seized upon mismatch theory to explore our supposed ongoing proximity with our Paleolithic ancestors. This is the idea that Yuval Noah Harari (2015) tried to popularize with his bestseller *Sapiens*, in which he first evokes a temporal proximity, reminding us that the last 10,000 years in which *Homo sapiens* has lived by working in the land and keeping livestock are merely the blink of an eye in comparison to the tens or even hundreds of thousands of years for which humans hunted and gathered. He also evokes a physiological and psychological proximity, which is the logical result of this temporal proximity:

> Even today, scholars in this field claim, our brains and minds are adapted to a life of hunting and gathering. Our eating habits, our conflicts and our

8 James C. Scott (2017) showed that Paleolithic hunter-gatherers had a wide portfolio of subsistence options, and that their skeletons showed far fewer markers of nutritional stress than those of the inhabitants of the earliest states. This suggests that they had more varied diets. However, we could hypothesize that the first ate more irregularly than the second, being more subject to the vagaries of the weather, to the presence or absence of game, to the practical difficulties storing food, etc.

> sexuality are all the result of the way our hunter-gatherer minds interact with our current post-industrial environment, with its mega-cities, aeroplanes, telephones and computers. This environment gives us more material resources and longer lives than those enjoyed by any previous generation, but it often makes us feel alienated, depressed and pressured. To understand why . . . we need to delve into the hunter-gatherer world that shaped us, the world that we subconsciously still inhabit.
>
> (Harari 2015, 55–56)

Therefore, in this book, we want to explore how the history of the human species and its physiology is mobilized to justify new practices and produce new forms of dietary reflexivity.

(De)politicizing everyday life

In what Victor Li (2006) calls the "neo-primitivist turn," there is the idea that, today, Paleolithic hunter-gatherers constitute models of life based on multiple inspirations, because for us, they are voiceless. Although paleo-anthropological knowledge is growing, theories about how these people lived are still hypothetical. Moreover, they represent humans before the state, before our "enslavement by priests, kings, and bosses" in the provocative words of John Zerzan (1994), one of the figures of the anarcho-primitivist movement. They are, in a way, a blank page onto which everyone can project their fears, desires and fantasies. This makes them an easily appropriable figure, which allows us to consider the plurality of political positions with regard to this enthusiasm for the wild and the rewilding movement. It is this plurality that we want to investigate in this book. Let us evoke some of the most contrasting positions.

On the one hand, rewilding can appear objectionable. One of the possible criticisms of mismatch theory, mentioned earlier, is that it peddles biological determinism: the idea supported by Harari that we still subconsciously inhabit the world of hunter-gatherers is limiting and de-politicizing. It essentializes our behaviors and supports a certain return to our instincts, against which the social sciences have been constructed. Although eating like our Paleolithic ancestors is currently promoted as a nutritional model, the idea of being inspired by their way of life is also connected to a moral question. It encourages a return to a life that is closer to nature, simpler and more authentic, guided by our instincts. This is also, implicitly, a more stereotyped life, because in our minds, we all have these images, reconstructed and disseminated by our schoolbooks and certain successful movies, of prehistoric humans wearing animal skins, living in caves and struggling every day to survive. They portray strong differentiation between the roles of men and women, which appear to be rooted "in nature": Men, being bigger and physically stronger, hunt large game, which they track for days, armed with spears and ingenuity. Women remain at home, looking after the children, keeping the fire burning, and sewing clothes. Today, thanks to gender archeology (Sørensen 2000), we know that this great founding narrative of our human condition is part fantasy and based on many errors of interpretation. It seems to us that the idea of taking inspiration from the Paleolithic way of life implicitly promotes a primitivist ideology that would contribute to the essentialization of gender roles and could work against social progress toward equality. The effect of this ideology would be to consolidate the idea that there are "masculine" and "feminine" foods and tasks that fall "naturally" to men and women. We find ourselves here at the root of the relationships of domination and the inequalities at the crossroads between gender and food (Devault 1991; Counihan and Kaplan 1998; Parsons 2015; Fournier et al. 2015).

On the other hand, the idea of rewilding our diet can easily be idealized, because it draws on the radical environmental movement and encourages a certain politicization of our choices and practices. This passion has a transformative potential. It could thus be interpreted as a kind of secularized asceticism (Sloterdijk 2013) in the form of reflexivity about diet, our bodies and the world we inhabit. The impoverishment of the self through the choice of a particular diet could in fact correspond to a more general ambition to control a primordial anthropological act without being guided by the abundance of consumer society. Food becomes a physical and spiritual exercise for getting as close as possible to an ideal of simplicity. This asceticism also involves learning new techniques, identifying local sources, developing new recipes, testing new flavors and becoming familiar with new sensations. As such, this voluntary rewilding of one's diet corresponds to an attempt to politicize daily life. This is not a formal politicization in the form of a vote or joining a social movement but a politicization through ordinary life and from below, a politicization of the smallest action (Pruvost 2021). This can be linked to Scott's *infra-politics* (1990) or the *ruses* evoked by Certeau, Giard, and Mayol (1998): adopting a "paleo diet," for example, could be a sidestep that falls within growing dietary reflexivity and that would in turn allow people to sharpen their critical sense and to distance themselves from the increasingly standardized, industrialized and globalized food offer. In this, the movement to rewild food could constitute a form of agency (Abbots 2017), a chance for everyday resistance and protest.

A *process of self-optimization*

The desire to move toward a more natural diet, and, through this, toward a simpler and more authentic life is somewhat paradoxically similar to a search for optimality: it is about being more reflexive, which means being more attentive to the potential impact of diet on one's body, health and environment. In a way, this fits the neoliberal logic and its core requirements for performance, control and individual responsibility. We believe that it is important to adopt a critical posture regarding this phenomenon, because by maintaining a competitive relationship with the world and with others, it can cause anxiety and guilt for those who do not have the material and symbolic resources required to participate. Dietary reflexivity is a long-term issue, which allows the most affluent to think about the future, but it generally traps the most impoverished in the urgent matters of everyday life. That said, this promise of rewilding food seems to attract relatively contrasting social profiles, as we will see in the different case studies in this book. Beyond this aspect, we want to draw attention to the processual dimension of this phenomenon, that is, the shift from rewilding one's diet to a possible rewilding of the self. Rewilding one's diet means questioning one's eating practices and their consequences, especially on the ecological, health and moral levels. Thus, eaters in search of self-rewilding express both a concern for the self and an ethical relationship to their daily practices. They engage in what we call a logic of "self-optimization" where "optimality is to be sought not so much in the more or less successful adoption of the imperative to maximize the use of the self as in the capacity of individuals to appropriate it, to find a satisfactory compromise" (Dalgalarrondo and Fournier 2019, 643).

Along the same lines, King et al. (2018) have shown that self-optimizing practices may equally take an aesthetic form, coming close to or even melding with a search for perfection. The authors distinguish two operational modes of this perfectionism. The first relates to a benchmark horizon that, as it were, guides the individual through an ascetic practice. The second takes the more specific form of a desire for enhancement and, in this sense, aligns with contemporary transhumanist philosophy (non-adapted/suboptimal body). In making the body central, self-optimization reveals a particular understanding of health and well-being, which is seen as a potential to be expressed. The collection of case studies presented in this book could help, in our view, to explain the fact that the first effect of self-optimization for those who practice it is one of detachment:

> We thus consider the processual dimension to be primordial, for it allows us to examine the effect of reflexivity inherent in it. The process is by definition non-linear: it is a construction in progress. Like a stream that finds its course by flowing with the topography, the process is adaptive, meandering and built up from contingencies. Because it involves *introspection*, a work of incorporation, it alters the way in which the individual is affected. Thus there is in the notion of process the possibility of slippage, a chance discovery, the establishment of a new relationship to society and its ideologies, to one's body and one's sensations.
>
> (Dalgalarrondo and Fournier 2019, 641)

In the case of rewilding, it seems that this process can give rise to different forms of exploration of the self and the world. One form focuses on the inner potential of each individual and is comparable to the flourishing domain of personal development: a physiological, sensory or even moral personal experience, which brings discovery through development and serendipity, and which fuels a certain egotism. A second form, which could be described as an exploration of the world, could arise out of the first: here, the rewilding movement allows us to envisage more distant, sometimes unexpected horizons, and possibilities of using diet and incorporation (Fischler 1988) to build new relationships with the world, other people, and other beings (particularly animals and plants). Perhaps, in the words of Hartmut Rosa (2019), this is a way of creating an "oasis of resonance" within our everyday lives that allows our relationship with the world to resonate. One of the challenges of the book is therefore to link these two processes: rewilding food and the self.

Structure of the book

The book is organized into three parts, each of which consists of conversations between social scientists. A first chapter, taken from fieldwork, is followed by a shorter commentary written by one or more researchers specializing in the theme in question. We hope that this original format will promote a "dialogue" between contrasting points of view and show the controversies at work. We have chosen to focus on European societies in an attempt to understand this appeal of the wild and the naturalness of food in geographical and political contexts where artifice, industrialization and ready-to-eat food have increased rapidly. The organization of the book allows us to navigate the three main themes that we feel are crucial to investigating the process of rewilding food and the self: 1) how the hunter-gatherer has become a resilient and inspiring model of life, 2) how practices aiming to reconnect with "natural processes" are developing in a market economy, and 3) how the references to the wild in food discourses and practices may challenge our identities. These three themes bring together ten case studies using qualitative and ethnographic methods, based on fieldwork conducted in Denmark, Finland, France, Italy, Norway and Switzerland.

Part 1: The taste for tradition and the hunter-gatherer model

This first part is dedicated to one of the practices that most embody the hunter-gatherer way of life and subsistence: hunting.[9] Running through modernity, there seems to be an idea of a necessary primitivist regression, of a saving degrowth, the objective being to get closer to a simpler, more eco-friendly and optimal way of life through wild food. We want to understand who, in sociological terms, is engaged in such a practice (said to be declining in Western societies) today and to identify the contexts in which game meat is consumed and the reasons why such a product is valued. The field surveys presented here emphasize that it is the authenticity of the practice and the qualities projected onto the hunting products that are pursued: because of their wildness and the imaginary it sparks, they are seen as more natural, healthier and better. They help to anchor eaters in a territory, to connect them to a culture, a past and an identity. The environmental dimension is also highly present in their discourses, even though it might seem that hunting, despite helping to maintain a close relationship with nature and the beings that live in it,

9 Foraging will be discussed in the third part of the book (to grasp the contemporary stakes of this practice in urban contexts) and in the conclusion (to consider its experiential rather than subsistence aspect).

could also pose concrete threats given the current biodiversity crisis. Conversely, killing animals is sometimes presented as a way of regulating so-called "invasive" species such as roe deer (*Capreolus capreolus*). Remember the leitmotif originally behind the rewilding movement: "let nature take its own course." Here, it is rather like humans want to take their place within the great food chain and thereby reintegrate themselves into these natural processes.

In the first conversation, Olivier Lepiller, Ophélie Roudelle and Sandrine Dury explore the consumption of meat from large game in the south of France, where wild boar, red deer and roe deer populations have boomed in recent years. Based on their qualitative survey, they construct an enlightening typology of eaters, which shows the many qualities projected onto wild meat. There is of course the search for authenticity and the need to be part of a cultural heritage requiring preservation, but there is also a strong promotion of game meat, because it comes from animals that have lived and wandered freely in their natural habitat, unlike those from intensive livestock farming. For (sometimes idealized) health and nutritional reasons, but also for moral reasons, therefore, large game meat could attract new groups, such as flexitarians, who prioritize quality over quantity and consider animal welfare. In his commentary, Sergio Dalla Bernardina looks at the opposition between domestic and wild, a symbolic construction that has been with us since the Neolithic, and which still seems highly persistent today. Although he considers that the naturalness attributed to game meat is an obvious reason for a renewed interest in consuming these products, he would not have imagined that a quasi-vegetarian minority (flexitarians) would one day start eating game for moral reasons. He emphasizes the strangeness of the situation, since this position can provide the defenders of meat eating with new arguments.

Part 2: Promises and market

This second part attempts to show how certain practices aiming to reconnect with "natural processes" are developing in a market economy. As we indicated earlier, wildness is a selling point. Here, we want to better understand the arguments used by the promoters of two recent products: natural wine and fasting retreats. In the first case, we try to show how a new artisanal winegrowing milieu is emerging, functioning, unfolding and becoming (almost) institutionalized. In the second case, we will see how the practice of fasting is reemerging in Western societies, (almost) stripped of its spiritual dimension in order to better fit the wellness and self-help market. In both the cases, for which the fieldwork was conducted in France and Switzerland, the promises are primarily health-related and moral. These products have something strange in common: both are commodified while also criticizing consumer society. This is even one of the arguments on which they are marketed and one of the reasons why consumers are drawn to them. They tie in with a critique of modern eating, symbolized by pesticide use and the industrialization of food. As such, their promoters encourage a return to nature, which is again idealized. There are numerous references to the wild: the aim is to go back in time in order to get as close as possible to the origins, when wild grapes grew freely and without artifice, and when hunter-gatherers effectively practiced intermittent fasting and lived in harmony with nature. The promise is therefore also that of self-exploration, and, more broadly, exploration of the world we have inherited and in which we want to live.

The second conversation is dedicated to the world of "natural" wines (those made without synthetic chemicals and with minimum intervention). Christelle Pineau has conducted an in-depth ethnographic study among these new winegrowers who want to give a kind of autonomy back to vines and a freedom of expression to wines. The theme of the wild is omnipresent in this environment. First, in the discourse of winegrowers, there is an anthropomorphization of vines and particularly of their wild grape ancestors, which demonstrates a desire for deeper exploration of the relationships between humans and plants. The aim is thus to give back as much freedom as possible to vines, one might say to rewild them, so that they can fully express their intrinsic qualities. The winegrowers try to take a back seat and not intrude. They are in service of the plant. For this reason, they are discredited by the dominant wine world, which sees them as uncivilized (i.e., wild)! Finally, the materiality of the wild can be found even in the glass, when indigenous yeasts and bacteria give the wine an animality. Léo Mariani's commentary starts by reminding us that the French term *nature* is ambiguous: as a noun, it designates an essence, while, as an adjective, it designates a pragmatic attribute. Thus, describing a wine as *nature* in French emphasizes its spontaneity more than its naturalness. So does the French term *nature*, translated here as *natural*, require inverted commas or not? What looks on the surface like an epistemological debate leads us to consider relationships between winegrowers and their products in a highly concrete manner. Because by choosing to preserve the autonomy of their wines, winegrowers who describe their wines as "*nature*" or "natural" allow them to express some of their own qualities. Temporarily adopting "the point of view of the wine" puts humans' point of view (and especially their role in rewilding wine) in perspective. Mariani tells us that "natural" wines are wild by definition.

The third conversation concerns fasting. It is not about the fasting promoted by many religions (Lent in Catholicism, Ramadan in Islam, etc.) but about so-called wellness fasting, a thriving practice in certain European countries, marketed in the standardized form of one-week "fasting and hiking" retreats. Sébastien Dalgalarrondo and Tristan Fournier study the promises of these retreats by examining the French market and particularly the discourses used by the retreat organizers. Two strands emerge from their analysis. The first is that of the animal model: promoters of fasting regularly remind us that all mammals fast, particularly when they are ill, and that human beings can go without food for around 40 days. This promotion of interspecies biology can be interpreted as a metaphorical invitation to rewild ourselves. The second strand is that of the primitive model: fasting is described as a chance to reconnect with our Paleolithic ancestors. Based on evolutionist theories, the modern individual is depicted as handicapped, degenerated and weakened by physiological maladaptations. A fasting retreat is therefore an experience, a scientific self-experiment, a biotechnological intervention "with bare hands" allowing the individual to optimize their body by reactivating their "true nature." In her commentary, Ophélie Bidet connects this self-exploration to forms of exploration of the world in a highly original manner. Having conducted a study among fasters in France and Switzerland, she shows that these retreats detach us from the social as much as they connect us to it. By breaking free from certain social conventions relating to diet, such as eating three meals a day at set times or the norm of breakfast, participants get a more concrete idea of how those suffering hunger feel and can therefore engage in forms of solidarity. In fact, she suggests that fasting is a way of constructing the self, as much as it helps to construct physical or symbolic links with others.

Part 3: The wild at the interstices: a route to empowerment?

The aim of this third part is to consider how this process of rewilding food and the self is expressed in post-modernity. The following case studies, for which the fieldwork was carried out in France, Italy, Finland and Norway, focus on two normative frameworks in which the rhetoric of the wild is mobilized in unexpected terms: urban life and the gender order. The city is a space that has historically been constructed in opposition to nature, but within which foraging and rummaging practices are now developing to feed, explore and optimize oneself. As for the gender order, which refers to the ways in which a society is organized around the roles and responsibilities of women and men, mobilizing the register of the wild to talk about food generally reinforces a "hegemonic masculinity" (Connell 2005) but may also allow us to circumvent it. In fact, here we will explore the ways in which these interstices create unexpected spaces of exchange and original confrontations of discourses and ways of life (Tsing 2015). Uprooted from its normative framework, the register of the wild thus loses its stereotyped nature, or conversely, plays on it. As we will see, this movement contributes to blurring the boundaries. It often appears disconcerting and sometimes transformative. Thus, the main objective is to understand the extent to which rewilding food and the self contributes to challenging our identities and to creating possible forms of emancipation.

The fourth conversation is devoted to urban foraging and scavenging practices. Flaminia Paddeu and Fabien Roussel explore the possibilities for capturing and reconnecting with nature in the city, and sometimes even eating it by foraging for plants in Greater Paris. Their research took them to the peri-urban margins: wastelands, vacant lots, edges, abandoned orchards or interstitial spaces, where people with often very different profiles and very different motivations come to forage, all of them taking advantage of recent policies in favor of biodiversity. In their chapter, urban foraging appears as an enjoyable way of rewilding the self, allowing participants to fully express their relationship to the city and the environment. This sometimes goes against the normative approach of the public authorities, which above all focuses on reconciling aesthetic and nature conservation objectives. The researchers examine the articulation between informal practices and regulatory policies in public spaces, and access to land and resources in the context of urban capitalism. Ultimately, foraging highlights the power relations at work between the various actors directly or indirectly involved in the management of green spaces. Sabine Mégarbané's commentary compares this practice to scavenging for food waste. Her ethnographic study on the freegan movement in France reveals numerous similarities with foraging. Here, the interstices are not forgotten green spaces but dumpsters: freegans rummage in these waste receptacles, put their heads inside them, or sometimes even climb into them, using all their senses to evaluate whether waste can become food again. This is presented as an instinctive process and can even go as far as justifying the consumption of expired food: by regularly exposing their bodies to bacteria, some freegans believe that they can stimulate their immune systems and perhaps optimize their physiology. Here, the conversation allows us to identify patterns shared by the two practices: voluntary wandering (to find edible plants or waste), the coproduction of the urban space and an original form of self-rewilding in environments that are, a priori, not very conducive to it.

The fifth and final conversation concerns mobilizations of the register of the wild in discourses on food, the body and gender. Nicoletta Diasio and Vulca Fidolini look at how men aged 40 to 60 in France and Italy use animal metaphors to describe their diet and manage growing older. Adopting a Foucauldian perspective, they see dietary practices as "technologies of the self," that is, procedures that allow individuals to maintain, adapt, and sometimes transform their identity according to the context. The midlife crisis that these men are experiencing is a time for dietary changes, because the body is undergoing transformation or because problems and pathologies are arising. These changes are generally accompanied by a renegotiation of gender identification processes. Thus, three avenues are analyzed. They sway between forms of association with and dissociation from the domestic and wild figures of animals, leading some to assume or even promote the profile of hegemonic masculinity, and others to reject it and therefore envisage a possible fracturing of the gender order. The commentary by Gun Roos reexamines a field study that she completed in Norway and Finland over 20 years ago, focusing on how male loggers and carpenters viewed the connections between their diet and their health. She takes us on a journey into the heart of masculinities, during which we discover a strong valorization of food as "fuel to work" and of the possibility of working in contact with nature, from which her interviewees sometimes took food (berries and game). In the conclusion, she asks what would happen if she did the same study today. What would it reveal about the connections between gender, nature and diet, in an era when most men have grown up in cities and are increasingly obliged to control their moral and bodily behaviors?

References

Abbots, Emma-Jayne. 2017. *The Agency of Eating. Mediation, Food and the Body*. London: Bloomsbury.

Barlösius, Eva. 2008. "Entre Est et Ouest. La moralisation de l'alimentation en Allemagne." In *Manger. Français, Européens et Américains face à l'alimentation*. Edited by Claude Fischler and Estelle Masson, 156–174. Paris: Odile Jacob.

Bauman, Zygmunt. 2017. *Retrotopia*. Cambridge: Polity Press.

Beardsworth, Alan. 2008. "Nostalgie et érosion des compétences alimentaires: Le vécu des Britanniques." In *Manger. Français, Européens et Américains face à l'alimentation*. Edited by Claude Fischler and Estelle Masson, 135–154. Paris: Odile Jacob.
Bourdieu Pierre. 1984 [1979]. *Distinction: A Social Critique of the Judgment of Taste*. Translated by Richard Nice. Cambridge, MA: Harvard University Press.
Carolan, Michael. 2015. "Re-Wilding Food Systems: Visceralities, Utopias, Pragmatism, and Practice." In *Food Utopias: An Invitation to a Food Dialogue*. Edited by Paul Stock, Christopher Rosin and Michael Carolan, 126–139. London: Routledge.
Certeau, Michel de, Luce Giard, and Pierre Mayol. 1998. *The Practice of Everyday Life. Volume 2: Living and Cooking*. Translated by Timothy J. Tomasik. Minneapolis, MN: University of Minnesota Press.
Connell, Raewyn. 2005. *Masculinities*. Cambridge: Polity Press.
Counihan, Carole, and Stephen L. Kaplan, eds. 1998. *Food and Gender. Identity and Power*. London and New York: Routledge.
Counihan, Carole, and Valeria Siniscalchi, eds. 2013. *Food Activism. Agency, Democracy and Economy*. London: Bloomsbury.
Dalgalarrondo, Sébastien, and Tristan Fournier. 2019. "Optimization as Morality, or the Routes of the Self." Translated by Rachel Gomme. *Ethnologie Française 176*(4): 639–651.
Dalgalarrondo, Sébastien, and Tristan Fournier. 2020. *L'utopie sauvage. Enquête sur notre irrépressible besoin de nature*. Paris: Les Arènes.
Dalla Bernardina, Sergio, ed. 2012. *L'appel du sauvage. Refaire le monde dans les bois*. Rennes: Presses Universitaires de Rennes.
Descola, Philippe. 2013 [2005]. *Beyond Nature and Culture*. Translated by Janet Lloyd. Chicago, IL: University of Chicago Press.
Devault, Marjorie. 1991. *Feeding the Family. The Social Organization of Caring as Gendered Work*. Chicago, IL: University of Chicago Press.
Fischler, Claude. 1988. "Food, Self and Identity." *Social Science Information 27*(2): 275–293.
Fischler, Claude and Estelle Masson, eds. 2008. *Manger. Français, Européens et Américains face à l'alimentation*. Paris: Odile Jacob.
Fournier, Tristan, Julie Jarty, Nathalie Lapeyre, and Priscille Touraille. 2015. "Alimentation, arme du genre." *Journal des Anthropologues 140–141*: 19–49.
Harari, Yuval Noah. 2015 [2011]. *Sapiens: A Brief History of Humankind*. New York: Harper.
King, Vera, Benigna Gerisch, Hartmut Rosa, Julia Schreiber, and Benedikt Salfeld, eds. 2018. *Lost in Perfection: Impacts of Optimisation on Culture and Psyche*. London: Routledge.
Knight, Christine. 2015. "We Can't Go Back a Hundred Million Years. Low-carbohydrate Dieters' Responses to Nutritional Primitivism." *Food, Culture & Society 18*(3): 441–461.
Lamine, Claire. 2005. "Settling Shared Uncertainties: Local Partnerships Between Producers and Consumers." *Sociologia Ruralis 45*(4): 324–345.
Lepiller, Olivier. 2012. "Naturel." In *Dictionnaire des cultures alimentaires*. Edited by Jean-Pierre Poulain, 905–911. Paris: Presses Universitaires de France.
Li, Victor. 2006. *The Neo-Primitivist Turn: Critical Reflections on Alterity, Culture, and Modernity*. Toronto: University of Toronto Press.
Loyer, Jessica, and Christine Knight. 2018. "Selling the 'Inca Superfood': Nutritional Primitivism in Superfoods Books and Maca Marketing." *Food, Culture & Society 21*(4): 449–467.
Luhmann, Niklas. 1979. *Trust and Power*. Translated by Howard Davis, John Raffan, and Kathryn Rooney. Chichester and New York: John Wiley & Sons.
Morin, Edgar. 1975. *L'esprit du temps 1. Névrose*. Paris: Grasset.

Parsons, Julie M. 2015. *Gender, Class, and Food: Families, Bodies and Health*. Basingstoke: Palgrave Macmillan.
Poulain, Jean-Pierre. 2009. *Sociologie de l'obésité*. Paris: Presses Universitaires de France.
Poulain, Jean-Pierre. 2017 [2002]. *The Sociology of Food. Eating and the Place of Food in Society*. Translated by Augusta Dörr. London and New York: Bloomsbury.
Pruvost, Geneviève. 2021. *Quotidien politique. Féminisme, écologie, subsistance*. L'horizon des possibles. Paris: La Découverte.
Rosa, Hartmut. 2019 [2016]. *Resonance: A Sociology of Our Relationship to the World*. Translated by James Wagner. Cambridge: Polity Press.
Scott, James C. 1990. *Domination and the Arts of Resistance: Hidden Transcripts*. New Haven, CT and London: Yale University Press.
Scott, James C. 2017. *Against the Grain: A Deep History of the Earliest States*. New Haven, CT, and London: Yale University Press.
Sébastia, Brigitte, ed. 2017. *Eating Traditional Food. Politics, Identity and Practices*. London: Routledge.
Shiva, Vandana. 2020. "Rewilding Food, Rewilding Farming." *The Ecologist*, January 24.
Siniscalchi, Valeria. 2020. *Slow Food: The Economy and Politics of a Global Movement*. London: Bloomsbury.
Sloterdijk, Peter. 2013. *You Must Change Your Life*. Translated by Wieland Hoban. Cambridge: Polity Press.
Sørensen, Marie Louise Stig. 2000. *Gender Archaeology*. Cambridge: Polity Press.
Tsing, Anna Lowenhaupt. 2015. *The Mushroom at the End of the World. On the Possibility of Life in Capitalist Ruins*. Princeton, NJ: Princeton University Press.
Zerzan, John. 1994. *Future Primitive and Other Essays*. New York: Autonomedia.

Part 1

The taste for tradition and the hunter-gatherer model

Conversation 1. Hunting and game consumption today

1 Eating wild game

A new carnivorous morality?[1]

Olivier Lepiller, Ophélie Roudelle and Sandrine Dury

> "Agricultural damage: hunters no longer want to be the only ones to pay."
> [translation from French] (journal, *Les Échos, 25 November 2021*)

> "Selling meat is bad."
> (hunter from the Occitanie region)[2]

> "For me, it's easier to justify eating wild game than vacuum-packed meat you buy in the supermarket."
> (an occasional eater of wild game, who is not a hunter)

What does eating wild game mean in France today? How is wild game acquired, dressed, cooked and shared? How does this practice fit into socio-cultural changes that affect meat consumption, food and eating in general and, even more than that, our relationship with wildlife? These are the questions that guided the investigation presented in this chapter.

Several major social and socio-ecological reconfigurations are intertwined around the subject of wild game in France (principally large game such as wild boar, roe deer, red deer and, to a lesser extent, fallow deer). For most French people, wild game is not a typical part of their diet, and in fact, its consumption has declined in recent decades. There is a presumption that it will progressively disappear from "edible space" (Poulain 2017), in other words, the general repertoire of foods commonly eaten and considered edible. This evolution is situated in a context marked by increasing urbanization and a demographical decline in the hunter population. Over the last 100 years, the urban portion of the population has grown from 50% in the 1920s to almost 80% in 2021, and in 1975, there were 2.2 million hunting licenses in the country, compared to 1.139 million in 2017 (Durand 2018).

1 Text translated from French by Ben Boswell.

2 Quoted by Lavergne et Faure, 2018. All direct speech quotations in this chapter are translations from French.

DOI: 10.4324/9781003243496-4

Over the last 30 years in France, meat consumption in general has declined slowly but regularly and is the object of growing concerns about its place in environmental, public health, ethical, political and identity issues (Interbev 2017; Bartoli, Hauser, and Prunier 2021). The various diets and eater philosophies that forgo meat (from simple vegetarians to completely vegan eaters) still represent a small minority but seem to be gaining terrain, particularly among the young. In a recent French questionnaire, 20% of the respondents identified themselves as flexitarians, a term coined to designate eaters who voluntarily reduce their consumption of meat for non-economic reasons (Avelin 2019). These individuals tend to prefer "less but better" and are drawn to more exacting production, notably in terms of animal well-being.

Overall, the meats from large wild game animals have followed this downward trend. They are counted in the category of "other meats," along with equid, rabbit, small game and edible offal, and their consumption has been in constant decline since the second half of the 1970s (SSP – FranceAgriMer 2020), falling from the carcass equivalent of 15 kilograms per year per inhabitant to two or three kilograms today. However, this drop in consumption can be questioned. In part because it doesn't distinguish between the subcategories but also because it is only calculated on meat that is commercially available. In a representative sample of the French population, Cazes-Valette (2006, 2007) found nearly 40% who never eat wild game. It is therefore quite probable that a small French minority, close to the hunting community, eat a large majority of the game consumed in the country. In fact, Jobard et al. (2016) found that 90% of the meat from wild boar and cervid hunting in France was consumed by the family and friends of the hunters themselves. At the same time, most of the game sold commercially in France (largely in the fall and winter seasons) is imported from northern or eastern Europe, New Zealand or Australia. Additionally, some of this imported meat comes from commercial game production rather than hunted wild game.

Nevertheless, the meat of large wild game is still imbued with a powerful symbolic image. It was long considered a privilege of the noble class in monarchical France. It progressed from a central position in medieval kitchens to the haute cuisine of the court, and finally found a position of honor in bourgeois gastronomy. Today, it is still marked by a certain prestige and can be given a central place in festive meals (family reunions, holidays).

In addition, there are contemporary social phenomena that can lead to the consumption of wild meat: growing criticism of industrial meat production and slaughterhouses, the search for a sense in the human sources of violence that are inherent in the ways we live and feed ourselves, the pressing need to establish new predation ethics in a world plundered by human activity, and the desire to reappropriate living conditions by recovering a sense of autonomy in an uncertain and anxiety-inducing world.

Accessibility to wild game meat is evolving. The hunting of small game birds and animals (rabbit, hare, partridge, pheasant, etc.), which were historically the target of small farmers and the working rural population, has declined, and their meat is becoming less accessible outside of commercial production and markets.

By contrast, the population of cervid and wild boar has been growing, and the working and middle classes are participating more and more in hunting these larger game animals which were historically the prerogative of the upper social classes (Stépanoff 2021). This evolution in participation has benefited from the increasing cost of crop damage inflicted by wild game. Legally, the French hunting federations are required to indemnify farmers for crops damaged by large wild game and therefore have a financial interest in managing the population of these animals by harvesting them. In certain French regions, the non-commercial circulation and self-consumption of game meat from the hunting community have been overwhelmed by the increase in the quantity of meat.

The question becomes how to make use of this surplus, especially since the simple killing and disposal of the animal is incompatible with centuries-old injunctions to eat an animal killed, to not waste the meat and to honor the animal for its contribution as a source of food. One way to avoid wasting the surplus meat would be to commercialize it. But this possibility collides with a hunting ethic that associates game meat with being wild, belonging to no one, and thus free of charge; it falls into an economy of gifts and counter-gifts. There is also the moral question of commercializing a wild good,[3] which, strictly speaking, is not produced even if human effort is required to obtain it. In other words, whether it should be extracted from the world of hunting, where social relationships between humans and wild creatures intermingle in a way that recognizes these creatures as being endowed with their own essence, an ability to exist and reproduce spontaneously, and an aptitude to resist predation. Regardless of this question, commercialization initiatives already exist. The National Federation of Hunters has registered a brand name, *Gibier de chasse – Chasseurs de France* [Wild Game – Hunters of France], and the local federation groups have recently expressed interest in the feasibility of commercial supply chains (FRC-IDF 2017; Lavergne and Faure 2018).

Where and how are large wild game meats situated in contemporary eating habits in France? To answer this question, we must consider several elements and the relationships between them.

- New opportunities for the commercialization of hunting surpluses *versus* the traditional non-commercial distribution.
- The increasing abundance of large wild game *versus* the historical prestige associated with meat that was reserved for the ruling class.
- The decline of the rural population *versus* new perspectives on the rural landscape.
- The decline in the number of hunters *versus* the increasing value attributed to the wilderness.
- Criticism of animal husbandry practices *versus* new ethical perspectives on predation.

3 A good in a socio-anthropological sense and an economic sense.

The investigation: eating large wild game in the French department of Hérault[4]

This chapter is based on a qualitative sociological study carried out between May and July 2018. The study incorporated 20 semi-structured interviews conducted with individuals that eat large wild game meat. Three main themes were covered in the interview: the consumption of this type of meat, the participant's relationship with game meat and hunting and their food habits in general. The interview addressed these themes by attempting to situate them in the biographical trajectory of each interviewee. These participants were all recruited in the department of Hérault in southern France and in accordance with the following requirements of diversification: urban/rural, young/old, men/women and hunters/non-hunters. There were several means of recruitment: invitations to participate in the study were posted in shops specialized in local products and the sale of wild game in Montpellier (the largest metropolitan area in Hérault) and in another community about 30 kilometers away; contacts were made at a conference titled "Feeding yourself with nature, the conference of wild locavores!" [translation from French] organized by the regional hunting association; and a "snow-ball" effect followed initial investigations meant to diversify the sample. The interviews were completely transcribed and then analyzed using thematic content analysis.

The questioning that guided the investigation is built on the framework of a theory of practices (Warde 2016; Lepiller and Cazes-Valette 2018). This theory privileges an approach through practices – food practices in this case – rather than through the individuals – the eaters in this case. These practices are defined as a group of concrete activities and meanings. They correspond to acts or ways of acting which carry certain interpretations, and these interpretations depend on collectively accepted standards that create judgments. They are also linked to the emotions that evolve from the objectives or intentions attached to the activity. The modalities of engagement that cause individuals to adopt a practice are diverse, and this diversity corresponds to the polymorphism of a given practice.

Our investigation reveals a food practice taking diverse forms. It is possible to situate those forms on a continuum stretching from a close proximity with the hunting community and significant experience with the consumption of large wild game meat to an exteriority far from the hunting network and few encounters with such consumption. Three large forms emerge. The first is found deep inside the hunting community and the great outdoors. It comes from a cultural heritage and familiarity and is mostly linked with the non-commercial circulation of game meat. The two other forms, less connected to the hunting community, revolve around the criticisms of domestic meat production. They are interesting in several respects: they suggest a potential future increase in the valuation of the relationship between food and wildness, and they fit into a more general context of

4 This investigation was conducted in the framework of the study, *GibSud de France*, funded by the *Féderation régionale des chasseurs d'Occitanie* [The Regional Federation of Hunters in Occitanie].

questions surrounding meat consumption, criticisms of industrial food production and the politicization of food. In this sense, eating wild food appears like a way to take better care of yourself and, at the same time, take a moral stance against the exploitation that dominates contemporary society.

Wild game as a familiar food

Three subtypes can be distinguished within the first form of food consumption.

In the first, eating large wild game is a regular habit. "It really doesn't have to be a particular occasion; it is just an opportunity to share" (Ms. Camone). Although large wild game is often served at special or holiday dinners like those at the end of the year, it is not limited to those events. "Some cuts of the meat will be kept for large gatherings, and others will be kept to eat when we are alone, even without a large table full of family or friends" (Ms. Nénard).

A place in the hunting community increases the possibility of regular access through non-commercial sources (self-consumption, gifts and trading). The meat is conserved or stocked (sometimes in dedicated freezers) to be available in periods outside of the hunting season, and a variety of species are hunted (e.g., wild boar, red deer and roe deer). Some of the interviewed hunters mention hunting trips further from home in the search for game that is relatively rare (red deer) or not found ("isard," a Pyrenean chamois) in the department of Hérault.

A strong hedonic dimension characterizes this first subtype. This priority given to pleasure naturally becomes part of friendly gatherings: "Sometimes I invite family members, my father-in-law and mother-in-law, all the people who don't hunt, but like [wild boar]" (Mr. Noyer). But the pleasure is also found on an everyday basis. When Mr. Quina was asked on what occasions he eats large game meat, he answered: "Anytime, just for the pleasure."

The comments reveal a hierarchy of meat taste, and that of wild game is clearly presented as superior[5]: "It is the best meat!" (Mr. Quina). "I find it much tastier than other meat" (Ms. Nénard). "For me, I like it because the taste is more pronounced, it's a strong taste!" (Mr. Erard). In terms of culinary preparations, stews are highly valued, as are grilled preparations of freshly slaughtered meat during the hunting season.

This "strong" taste of "game," of "wild" meat (the expressions used frequently) is sometimes associated with virility. Mr. Dalais explained that the wild game meats are "rather strong and not compatible with all palates," before adding, "among guys I think it is ok, but when it comes to the women, we are dealing with things a little more sensitive." Moreover, this first subtype also has an affinity with the masculinity inherent in the hunting community itself.[6]

5 This superiority of wild meat over domestic meat is rooted in the myth of wildness and "black blood" highlighted by Bertrand Hell (1994) in his ethnological study of hunting in Europe.

6 The overwhelming majority of hunters are men. According to France's National Federation of Hunters, only 2.5% of its members who practice hunting are women: 25,750 out of 1.03 million

The frequent consumption of meat seems to be completely assumed, along with the killing it implies. This affirmation is evidenced by self-designation with the pejorative term, "viandard."[7] Although applied here to the eater and the hunter, this word was coined to indicate hunters with less than scrupulous morals who disregard the rules and hunt simply for the pleasure of killing as many animals as possible. Interviewees used the term as a gesture to reverse the stigma: "We are the viandards," declared Ms. Camone. And Mr. Dalais claimed to think of himself as "a good viandard." In a slight variation, Mr. Quina saw himself, more simply, as a "carnivore."

In a second subtype, large wild game meats do not always appear to be valorized for their specific qualities. By contrast, they tend to be seen as commonplace. As a hunter's partner expressed it, "For me it is just a meat like any other when it comes down to it. . . . I eat it like I eat any other kind of meat" (Ms. Begonia). Even a certain banality can be expressed: "I'm beginning to get tired of [wild boar] stew . . . it's getting a little predictable . . . you need to change the dishes and the way you cook it" (Ms. Béguin). Ms. Camelot, whose son-in-law is a hunter, expressed a modest taste for wild game in contrast to the valorization of the meat's strong taste in the first subtype: "Sometimes it can be a little too strong in terms of taste. You may need to tone down the taste. . . . I find the stews a little heavy."

This second subtype has a privileged place in the immediate social periphery of the hunters, thus allowing a degree of non-commercial access to wild game. When the social distance from the hunting community is greater (more extended family or friends) or if it is overextended (e.g., after moving), access can sometimes be found in commercial outlets. For example, this was the case for Ms. Géranine:

> When we visited the Lorraine [northeastern region of France], I went to a wholesale butcher, and he had wild game during the holidays. . . . Somewhat for the tradition . . . and to please my husband because I know he loves game.

This comment also illustrates the gender and nourishing dimension of this subtype. In our study, the individuals most identified with this subtype were women, mainly represented as the wives and partners of hunters. The game meat in these cases is cooked, above all, for others. Cooking appears to be a very important element for these women in caring for the couple and the household. It is a means of fulfilling the role that is expected of them, that of caring for others (Fournier et al. 2015).

Although this second subtype has some non-commercial access to game meat, that access is still not always quite as easy as in the first subtype. The less distinct link to the meat means that the consumption may be reduced to the hunting

---

practicing members. www.chasseurdefrance.com/decouvrir/les-chasseurs-aujourdhui/ (consulted on 2 September 2021).

7 "Viandard" is formed from the root "viande" (meat) and the suffix "-ard" which is found in many pejorative terms in French.

season or holiday meals and festive occasions when guests are entertained. Opportunities offered by the commercial environment can also play a role in prompting consumption, but those opportunities vary from region to region. In her comments, Ms. Géranine added a footnote: "I won't be going around Montpellier trying to find game meat." It is more difficult to find in Montpellier as compared to northeastern France where there is a tradition of commercializing game meat. The comments by Ms. Camelot confirm the importance of the commercial environment:

> If I see it [game meat] regularly, maybe at some moment I will buy it. . . . When I am at the butcher shop, if I see it, I may say to myself "I think I'll take that today."

The second subtype is very dependent on factors outside of the individual. These are found among the circle of family and acquaintances, especially in expectations related to the role of providing food. Although the strong wild taste of game meat is not generally valorized, it is appreciated among those within this nurtured social environment. Consequently, this subtype is subject to being easily transformed, notably in function of the evolution in links to the hunting community, family and marital relationships and the commercial environment. This evolution can trend toward less consumption, as in the case of Ms. Géranine, who eats game meat less frequently since she moved to Montpellier, or toward an increase, as in the interesting case of Mrs. Béguin.

After a period during her youth when she did not eat game meats, Mrs. Béguin began to change her diet when she became involved with a man who she belatedly learned was a hunter. "That was a shock," she said laughingly. She had thought of herself as "anti-hunting," but discussions with her partner lead to an evolution in her point of view. This new perspective and the access to wild game meat granted by her relationship led to its inclusion in her diet, "even if [I am] not ready to eat nothing but that."

In the third subtype, eating large wild game is above all the signification of a cultural identity. This type of consumption is not as frequent as in the first subtype because of limited proximity to the hunting community, and access to game through non-commercial sources is complicated and irregular.

The case of Mr. Castana illustrates well this subtype. Originally from Corsica, he has never been a hunter himself, but he grew up in a culture with hunters where large game meat was often eaten. His strong attachment to this type of meat is specifically connected to his identity:

> I associate wild boar with the village and family. . . . This is not my "madeleine de Proust" but almost. . . . It is truly associated with a cultural heritage, with the memories that can be almost nostalgic.

Today he lives in Montpellier and eats wild game rarely, compared to his youth in Corsica, and even less so, because he only eats game coming from his native island. His access is limited to when he goes there on vacation, or when he obtains meat from his network of family and friends: "For example, it's the first cousin of my brother-in-law who gives it to my sister, who gives it to me. Or it can also be a hunter friend who gives it to me. But it's not from commercial sources."

The culinary preparations are part of a heritage that is celebrated through the sharing of a meal, like at family reunions and get togethers.

Interviewer: When you have large wild game, how do you cook it?

Mr. C: That depends, if it is the leg of a young animal, for example, that can be roasted, but after marinating it. If it is a little older, it will need to be in a sauce, always with a little alcohol added . . . like a stew except it doesn't have the rind. There are no olives, just an onion, garlic, mushrooms, and the bay leaf.

Interviewer: And are you the one who cooks the game?

Mr. C: It's either me or my sister, but not my wife.

Int.: Ok, and your sister, does she cook it like you?

Mr. C: Ah, I'd say its ancestral, (laughs) so yes! So, well yes, but since she is older than me, I'm the one who cooks like her. She cooks like my mother, my mother cooks like . . . and so on.

In the case of Mr. Castana, no game meat was acquired commercially. In other cases, when non-commercial access was impossible (e.g., because of geographical or social distance), the meat could also be purchased from commercial sources to satisfy the ritual traditions of a family holiday meal.

This dimension of personal identity is shared by all three subtypes of the first form of food consumption. The third subtype just described shows that this dimension also concerns individuals who continue to cultivate roots in the hunting community even though the course of their lives has separated them geographically.

Mr. Castana's attachment to his native island is reflected in the incorporation of native game in his food habits. This testifies to the force with which food practices can trace the contours of our sense of identity (Fischler 1988). The regional culture of Mr. Castana's Corsican origins is distinctive. The mountainous island of Corsica is an example of socio-nature where cultivation is interlaced among wild spaces commonly known as the maquis.[8] The famous mountain pigs in Corsica are bred in liberty and celebrated for their propensity to go wild and reproduce with wild boars. The wild aspects of Corsican sociocultural identity correspond to that of the island which has been part of the stereotypical discourse on the Corsican population for several centuries (Franzini 2004). It is sometimes brandished as a standard by the Corsicans themselves as support for the spirit of insubordination

8 Maquis is the name for Mediterranean scrub forests of low and dense growth in the crystalline soil characteristic of Corsica. The French expression "prendre le maquis" (take to the maquis) means to hide from or join the resistance against authority. The French use "maquis" to designate the wooded and sparsely populated areas used in the resistance against German occupation during WWII.

toward the French mainland. Certain members of the pro-independence movement used to film themselves at night, hooded, in the middle of the maquis.

Wild game as a healthy food

In the second form of dietary practices, health benefits are the prevailing reason for eating the meat of large wild game. This form can be found among individuals close to the hunting community, including the hunters themselves and individuals outside of the community that only purchase game meat from commercial sources. The latter appear to be interested in the establishment of a commercial supply chain for local game, notably the shops specialized in local products, the open-air markets or the butcher shops, all places where they are used to shopping: "I am trying to shop closer to the producers . . . the most local and most direct possible" (M. Lilas).

A nutritional aspect is applied to game meats to justify their "healthy" quality:

> [The taste] is not what I am looking for in meat. . . . I think game meat tastes very good, but its taste is not the reason that I prefer it over other meats . . . on a nutritional scale, game meat is actually pretty good. . . . The amount of protein and also the amount of fat, there is less fat. And [wild boar meat] has more nutritive elements than classic pork meat.
>
> (Mrs. Myrte)

The predominance of this focus on health is also expressed in the aspect of being "natural." Wild game meats are seen as being the opposite of meat from farm-raised animals:

> The meat is much healthier than meat from farm-raised animals . . . these are animals that have eaten natural food . . . the wild animals that were not raised in a cage, that haven't been subjected to medical treatments.
>
> (Mr. Cotier)

This second form maneuvers among food practices that are largely focused on the attention to health, notably in terms of organic food. Mr. Géranine provides an example: "More and more, I am trying to prioritize organic products . . . to avoid those treated with pesticides, to have a better-quality product, to eat healthier." For Mr. Géranine, the meat's wild origin guarantees a degree of freedom from human intervention that can be harmful to health, like organic production guarantees the absence of dangerous synthetic chemicals. From his point of view, organic and wild share a common identity.[9] This shared identity of wild and organic can also be seen in the absence of "forcing" in food that is naturally

9 This link between organic and wild goes beyond cultural borders. During a recent mission in Gabon to study the consumption of wild game in a forested zone, several of our contacts jokingly qualified

grown or raised (Lepiller 2012). Forcing is a typical example of the "exploitation of nature" (Stépanoff 2021), a purely utilitarian perspective in the raising and cultivation of living beings, aimed at maximizing profit, sometimes at the risk of the eaters' health.

The case of Mr. Ulmo suggests that the emphasis on health can also become predominant following health issues that modify the individual's relationship to meat in general and particularly that of wild game. For several years, Mr. Ulmo has been diagnosed with a risk of developing gout because of high uric acid levels.[10] He was ultimately confronted with his first crisis shortly before the interview and now regrets his past excesses:

> It is related to the fact that I always refused to throw out game meat, so I always ate it until it was gone and obviously consumed too much. . . . And above all, I lacked information. I mean, what are the . . . the comparison between meat from hunting compared to industrial meats, in the level of uric acid. What are the conditions that cause it to increase? What are the . . . so, uh . . . there I regret it, so there, if I had known all that a little earlier, maybe I would have managed my way of consuming and . . . differently, as a function of those parameters. And I have underestimated the problem.
>
> (Mr. Ulmo)

Being particularly attentive, and caring for their health, the interviewees with a profile corresponding to this second form don't generally ignore the health alerts regarding the meat of large wild game, particularly those concerning Trichinella in wild boar meat and the risk of exposure to heavy metals. But these alerts are relativized by a sense of controlling the risks (Figuié et Fournier 2008; Fournier 2009). To be effective against the risk of trichinosis, Mr. Cotier noted, "we know that you have to cook the meat to more than 65 degrees [centigrade] at its center." Mr. Ulmo relativized the exposure to lead in his dietary past:

> I think that is for those who eat a lot, really a lot; I don't think I eat that much; I think I eat a lot, but not beyond a certain limit, even so! I think there are hunters who eat much more than I do, so the lead might be a problem for them. And in my opinion, the lead is a problem for those who especially eat small wild game. Because, well, the density of lead is concentrated in a small quantity, especially the thrush, etcetera. On a wild boar, that does what.

their wild game meat as "organic" like the input-free crops surrounding their villages (plantain, manioc, eggplant, peppers, etc.).

10 Gout is a pathology that provokes joint pain due to the accumulation of urate crystals in the joint. Uric acid is produced by metabolized purines, nitrogenous molecules whose principal dietary sources are meats. A kidney problem can cause excess uric acid and can be aggravated by this excess. Gout was a frequent pathology in nineteenth-century men from the bourgeoisie. Aging, an overabundant diet and sedentariness are known risk factors. Wild game meats are particularly rich in purines and their consumption is not recommended for gout sufferers.

> There isn't . . . there are three lead balls in 60 kilos of meat. I don't think that is dramatic in terms of eating lead. But ok I know the problem exists.[11]
>
> (Mr. Ulmo)

In general terms, this second form demonstrates a will to moderate the quantity of meat consumed, often by prioritizing better quality over quantity. This better quality is seen in opposition to industrial livestock farming. It can also be associated with domestic meat identified as a specific quality and perceived as non-industrial (small farm production, range fed, and labels like organic, or Label Rouge,[12] etc.).

Wild game as an ethical food

The third form of food practices is found in an ethic centered around "less but better" consumption of meat. This dimension was also part of the previous form with which it shares several points (this explains the overlap among participants illustrated as examples). However, in these cases, the consumption is guided primarily by ethical goals. This ethical dimension is about the hunting practices, the relationship with the environment and the relationship with animals.

For Mrs. Myrte, "behind the fact of killing the animal itself, it is respected and given its chance. . . . Hunting is more respectful of the animal than livestock farming." For Mr. Ulmo, all hunting should be limited to providing food if it is to remain ethical:

> It is out of the question to go hunting if it isn't to eat what you have killed. It loses all sense; the hunt has no more meaning . . . going hunting in some new place just to kill game for the prestige; I think that's unhealthy.

The notion of "natural" is mentioned, as it was in the previous form, but this time to characterize a moral relationship with the animal and the environment. The words of Mrs. Myrte describe two aspects of "natural" that allow her to

11 In 2018, a report by the French Agency for Food Safety advised limiting the consumption of wild game meat to three times a year. It also advised against any consumption by women of childbearing age because of the risk of exposure to pollutants such as lead (ANSES 2018). The methodology of this report has been criticized. The agency asked the hunting association to furnish pieces of meat for analysis. It is probable that hunters furnished pieces they would not have eaten, because it included the trajectory of the projectile that contains lead. The results could also be biased, overestimating the risk of exposure. However, Mr. Ulmo seems to ignore the environment that can be another source of heavy metal contamination in wild game, particularly in polluted areas (old industrial sites or mines, hunted wetlands, etc.). Rooting animals such as wild boar, a major quarry for hunters in France, are particularly exposed to this contamination, and the risk, although differing from one area to another, it is not measured. Given this large variable and the absence of more precise localized measurements, the agency obviously preferred to err on the side of caution.

12 Label Rouge is a French label for food quality that attests to a quality superior to other similar products. It is used on different products with varying specifications and can also be applied to crop and livestock farming methods.

mentally correlate the health of the person and that of the environment. In her mind, large wild game meat is included in a diet "that is healthy for me but also respectful to the environment; it isn't something industrial." Mr. Killy believes that hunted meats, contrary to those produced by the practices common to animal husbandry and slaughterhouses, allow respect for the animals: "I want the animal to be respected from A to Z, without any suffering."

This third form has found a privileged place in the hunting community. Food as a justification for the act of hunting and killing – hunting only what you can eat or give to someone else to eat and, above all, not throw away – corresponds to the ethic of certain hunters like M. Ulmo. In addition, this form goes hand in hand with concerns for the environment that are shared by numerous hunters (Stépanoff 2021). This aspect is being used more frequently by hunting associations as part of an effort to "greenify" their activity, which is often the target of ecological organizations. In this spirit, the National Federation of Hunters launched a 2018 campaign to portray hunters as France's premier ecologists.

This form of large game meat consumption can also find adepts outside of the hunting community. Here, the case of Mr. Lilas is instructive. He is not a hunter and grew up in an urban setting without any direct contact with the world of hunting. His parents, "not at all hunters, . . . or farmers or anything like that," did, however, receive occasional gifts of meat from hunter friends, which gave him the occasion to eat large game meat from time to time. Sensitive to the animal rights movement, he went through a period when he did not eat commercial meat because of criticisms about the treatment of domestic livestock and mistrust related to health risks associated with antibiotics and other pharmaceutical products used in meat production. During this time, he did occasionally eat game meats because he was confident of their quality. When the interview was conducted, Mr. Lilas was no longer completely excluding domestic meat from his diet. He does moderate his meat consumption, restricting it to meat from a social network of livestock farmers that he found while living for a while on a farm. With respect to wild game, which he eats occasionally, he suggested that it is more natural:

> [It's] more natural to hunt your own meat rather than to have it right in front of you [in the grocery aisles], . . . there is a hypocritical aspect of buying meat in a supermarket; you are not face to face with the animal; you have this stuff wrapped in plastic; you buy it and there you are. When you hunt for meat, you first need to have the courage to kill it; personally, I don't know if I could . . . [but] for me, it is easier to justify eating hunted meat than to eat something wrapped in plastic and bought in the supermarket.
>
> (Mr. Lilas)

Mrs. Magnole is another instructive case. She identifies herself as being "vegetarian" for around 40 years, while conceding a certain degree of flexibility: "I haven't eaten meat outside of special occasions etcetera, but I am not a model in that field." When she does eat meat, it is for the most part large wild game, like the last time she ate meat: "I ate some last year because we invited the family, and there were twenty or so people, and in fact we were given some wild boar, so we cooked leg of wild boar – a seven-hour wild boar." In the spirit of this vegetarian ethic where she affirms wanting to stop "collaborating in the massacre," the meat of hunted animals has a different status: "I prefer wild game that is hunted and eaten over animals raised in a totally inhumane fashion and fed whatever." Living in a rural setting, coming from a family of hunters and knowing the overpopulation problem of wild boar, she understands the reasoning of a supply chain for the commercialization of wild game. She characterized the idea for one in the neighboring department (the Gard) as "great."

The wild road to salvation for meat eaters?

The first dietary form is organized around well-recognized dimensions of wild game consumption: these hedonistic-, gender- and identity-related aspects are linked with the practice of hunting and its close relationship with nature. However, the second and third forms suggest new perspectives that valorize large wild game in ways not entirely tied to the hunting community and its non-commercial circulation. These forms could thus be receptive to commercial supply systems that would involve new relationships between meat eaters outside of the hunting community, the hunters themselves and the wild domain. To understand these new perspectives, we need to situate them with respect to the larger social phenomena.

First, we can look at the nutritionalization of contemporary food (Dixon and Banwell 2004; Coveney 2006; Scrinis 2008; Poulain 2009, 2017; Fournier 2011; Lepiller 2012; Kimura 2013; Fournier and Poulain 2018). This process corresponds to a change in the hierarchy and purpose of eating that favors health as the goal that supersedes all others (repletion, sharing, conviviality, pleasure, etc.). It is illustrated by a preventive approach to health in which diet occupies an important place, along with the necessary diffusion of dietary and nutritional categories and knowledge among the population.

We have seen the arguments made by certain investigations asserting the superior nutritional quality of wild meat. The hunting associations themselves have participated in the process by producing and diffusing nutritional information about wild game. The National Federation of Hunters ordered and published a comparative study between hunted wild game and domestic meats which was conducted under the scientific direction of a hospital professor specialized in endocrinology, diabetes and nutrition (Ducluzeau 2004). This study emphasized the relative low lipid content of game meat and the high levels of micronutrients of interest such as phosphorus and iron. To give an example of this study's reception, we quote the title of an article from a regional journal based in Lyon: "Wild game meat reconciles gastronomy with dietetics"[13] [translation].

13 Le Progrès, 10 October 2017, www.leprogres.fr/jura/2017/10/10/la-viande-de-gibier-reconcilie-la-gastronomie-et-la-dietetique.

Additionally, these meats are valorized by the "paleo" or "prehistoric" diets that use nutritional justification and arguments borrowed from contemporary science to construct a historical narrative of the human diet. Their proponents argue that the hunter-gatherers were healthier than sedentary agricultural societies with cereal-based diets. In their view, a diet oriented toward small fruits, tubers, wild fish and game should be more adapted to human physiology than the diets based on agriculture that developed at the beginning of the Neolithic period.

But nutritionalization not only is about the dietary quality of food, but it also involves food safety. We have noted that wild game meat is sometimes proposed as way to avoid animal husbandry inputs (antibiotics and hormones in particular). However, there are also arguments pointing to the dangers to eater health from hunted game meat, such as concerns related to the presence of heavy metals like lead from hunting ammunition or from the environment. They can also be potential vectors for parasitic diseases, although appropriate cooking methods can eliminate that risk.

Characterizations of the health aspects of eating hunted wild game meats are mixed, but in a commercial framework, the arguments against their consumption could be countered by technical means (e.g., a traceability system, systematic analysis of meat on the market and information on adapted cooking methods).

The second social phenomenon is the "deanimalization" of diets. This term refers to a process, practically and symbolically observable, where food is disassociated with animality (Lepiller, Fourat, and Mognard 2016: Fourat and Lepiller 2017). Following are several dimensions of this process:

- the relatively more important decrease in the consumption of animal-based foods with a strong image of animality and blood (e.g., red meat, offal and wild game),
- the increase in "sarcophagous" meat processing that reduces the extent it resembles flesh or the animal itself and a decrease in "zoophagous" food (when the living animal is recognizable, such as roasted suckling pig) (Vialles 1988),
- the invisibilization of meat organicity, favorized by industrialization and commercialization of production systems (Bruckert 2021),
- the distancing from the killing process, beginning in the nineteenth century with the installation of slaughter facilities outside of cities (Vialles 1995), and
- contemporary animal activism, often linked with veganism, which denounces the dominant carnivorous ideology and the livestock farming and slaughterhouse conditions (Dubreuil 2013; Mouret 2016).

Since hunted game meats have a particularly strong sense of animality, and are thus stigmatized in the arena of deanimalization, how could their consumption benefit from this process? Shouldn't these meats be the first victims of any distancing from animality? Our study suggests the plausibility of an inverse hypothesis. This hypothesis describes a sort of counter-current in the heart of the strong flow of deanimalization: a sub-product traveling in the opposite direction. More precisely, it is the wild character of hunted game meats that could encourage their place among eaters who are searching for alternatives to the dominant relationship between meat and the exploitation of nature. This character could provide a new moral anchor for meat eaters and a new ethic toward predation.

Indeed, eating hunted wild game does offer a different way to think about the lives of animals and to put that thinking into practice. From this alternative perspective, animals have the leisure to live freely in the wild as nature intended, and the practice of hunting and killing, in contrast to industrialized slaughter, even leaves the animal "a chance to escape." In this way, hunted game meat might integrate into a food model we might call "silvatarianism" (following the model of pescetarianism) that would exclude animal meat except for that of hunted wild animals.

These types of practices resonate with contemporary reevaluations of wildness (Dalgalarrondo and Fournier 2020) and predation (Dalla Bernardina 2011). They are also part of a willingness to "not look away" from the killing implicated in all types of meat eating. Today, meat-eaters are publicly confronted with operations conducted by activists such as the French vegan association, L214, which often use films to divulge gruesome conditions in livestock farming and slaughterhouses. These acts call for people to assume responsibility for the killing, particularly urban eaters who are completely isolated from the experience of raising and slaughtering animals and cutting up their bodies. As an example of the reaction to this activism, we can cite the case of Facebook's Mark Zuckerberg, who made a public resolution for the year 2011:

> This year, my personal challenge is around being thankful for the food I have to eat. I think that many people forget that a living being has to die for you to eat meat, so my goal revolves around not letting myself forget that and being thankful for what I have. This year I've basically become a vegetarian since the only meat I'm eating is from animals I've killed myself.[14]

We can also cite the 2016 book of journalist Louise Gray, *The Ethical Carnivore. My Year Killing to Eat*, in which the author relates and analyzes a challenge to that of Zuckerberg's.

The abilities to not look away and to associate hunted animals with the paradoxical relationships of violence, identification and respect are precisely at the heart of the practice and ethics of hunting (Stépanoff 2021). In this sense, predation in the wild offers a moral framework, which urban eaters who are neither hunters nor livestock farmers can draw upon to develop new ethics of predation. This new ethic is not necessarily engaged directly by the eaters. It can be indirectly delegated to hunters capable of not looking away. In the sharing of the moral work of killing, the hunters would occupy the place of slaughterhouse employees, whose working

14 Quoted by the journal, *Business Insider*, 26 May 2011: www.businessinsider.com/mark-zuckerberg-is-only-eating-food-he-kills-2011-5.

conditions have been denounced along with the conditions in which the slaughtering takes place. New forms of solidarity become possible, connecting hunters and the rural world with urban eaters. But the conditions that would encourage their development through commercial relationships are hampered by complications previously mentioned, such as the moral concerns of selling wild game animals that belong to no one. However, existing socio-economic forms, such as short supply-chains and community-supported agriculture (or, in this case, community-supported hunting), could provide models for these new forms of solidarity.

Beyond reflections on the killing required, meat consumption reduced to that from hunted wild animals can be seen as the refusal of any responsibility to raise and keep animals to be killed for food, including small production, "traditional," "artisanal" or "cottage" industries. In this perspective, the practice rejects any justification for the consumption of domestic meat through the system of gift and counter-gift that traditionally supports the legitimacy of animal husbandry: a gift of life and care to the animals raised, and a counter-gift of death to feed the eaters' life (Mouret 2012). This foundation of the "domestic contract" (Larrère and Larrère 1997) is thus called into question.

References

ANSES. 2018. *Avis de l'ANSES relatif au " risque sanitaire lié à la consommation de gibier au regard des contaminants chimiques environnementaux dioxines, polychlorobiphényles (PCB), cadmium et plomb."* Agence nationale de sécurité sanitaire de l'alimentation, de l'environnement et du travail. www.anses.fr/fr/system/files/ERCA2015SA0109.pdf

Avelin, Christine, ed. 2019. *Combien de végétariens en Europe ? Synthèse des résultats à partir de l'étude "Panorama de la consommation végétarienne en Europe", réalisée par le CREDOC pour FranceAgriMer et l'OCHA en 2018.* FranceAgriMer. www.franceagrimer.fr/fam/content/download/62309/document/11_Synth%C3%A8se%20Panorama%20v%C3%A9g%C3%A9tarisme%20en%20Europe.pdf?version=1

Bartoli, Pierre-Hadrien, Morgane Hauser, and Anaïs Prunier. 2021. *La consommation de viande : Quelles nouvelles tendances?* Harris Interactive, Réseau Action Climat. http://harris-interactive.fr/wp-content/uploads/sites/6/2021/02/Rapport-Harris-La-consommation-de-viande-Reseau-action-climat.pdf

Bruckert, Michaël. 2021. "Chicken Politics Agrifood Capitalism, Anxious Bodies, and the New Meanings of Chicken Meat in India." *Gastronomica 21*(2): 33-46. https://doi.org/10.1525/gfc.2021.21.2.33

Cazes-Valette, Geneviève. 2006. *Le rapport à la viande chez le mangeur français contemporain. Rapport d'étude ethnologique.* Toulouse: Groupe ESC/CCIT. www.agrireseau.net/bovinsboucherie/documents/rapport%20%C3%A9tude%20ethnologique.pdf

Cazes-Valette, Geneviève. 2007. "Contre la viande, tout contre . . ." In *L'homme, le mangeur, l'animal qui nourrit l'autre?* Edited by Jean-Pierre Poulain, 159-171. Paris: OCHA.

Coveney, John. 2006. *Food, Morals, And Meaning: The Pleasure And Anxiety of Eating* (2nd Revised edition). London: Routledge.

Dalgalarrondo, Sébastien, and Tristan Fournier. 2020. *L'utopie sauvage. Enquête sur notre irrépressible besoin de nature.* Paris: Les Arènes.

Dalla Bernardina, Sergio. 2011. *Le retour du prédateur. Mises en scène du sauvage dans la société post-rurale.* Rennes: Presses Universitaires Rennes.

Dixon, Jane, and Cathy Banwell. 2004. "Re-embedding Trust: Unravelling the Construction of Modern Diets." *Critical Public Health 14*(2): 117-131. https://doi.org/10.1080/09581590410001725364

Dubreuil, Catherine-Marie. 2013. *Libération animale et végétarisation du monde: Ethnologie de l'antispécisme français*. Paris: Éditions du CTHS.

Ducluzeau, Pierre-Henri. 2004. *Dossier exclusif chasse et nutrition*. Fédération nationale des chasseurs. www.gibier-de-chasse.com/assets/telechargements/Etude_Ducluzeau.pdf?PHPSESSID=3nnponmbkjghoi10okj8l49192

Durand, Anne-Aël. 2018. "Sur 1,1 million de chasseurs, moins de 10 % possèdent un permis national." *Le Monde.fr*. www.lemonde.fr/les-decodeurs/article/2018/08/29/sur-1-1-million-de-chasseurs-moins-de-10-possedent-un-permis-national_5347594_4355770.html

Figuié, Muriel, and Tristan Fournier. 2008. "Avian Influenza in Vietnam: Chicken-Hearted Consumers?" *Risk Analysis 28*(2): 441-451.

Fischler, Claude. 1988. "Food, Self and Identity." *Social Science Information 27*(2): 275-292. https://doi.org/10.1177/053901888027002005

Fourat, Estelle, and Olivier Lepiller. 2017. "Forms of Food Transition: Sociocultural Factors Limiting the Diets' Animalisation in France and India." *Sociologia Ruralis 57*(1): 41-63. https://doi.org/10.1111/soru.12114

Fournier, Tristan. 2009. "Coping With New Food-related Risks." *Anthropology of Food, Varia*. https://journals.openedition.org/aof/5303?gathStatIcon=true&lang=en

Fournier, Tristan. 2011. *Une sociologie de la décision alimentaire. L'observance diététique chez des mangeurs hypercholestérolémiques*. PhD thesis in Sociology, Université de Toulouse II-Le Mirail, Toulouse.

Fournier, Tristan, Julie Jarty, Nathalie Lapeyre, and Priscille Touraille. 2015. "L'alimentation, arme du genre". *Journal des Anthropologues 140-141*: 19-49.

Fournier, Tristan, and Jean-Pierre Poulain. 2018. "Eating According to One's Genes? Exploring the French Public's Understanding of and Reactions to Personalized Nutrition." *Qualitative Health Research 28*(14): 2195-2207. https://doi.org/10.1177/1049732318793417

Franzini, Antoine. 2004. "Un peuple libre, sauvage et vertueux : Nature et politique dans la Corse du Quattrocento." *Médiévales. Langues, Textes, Histoire* 47: 63-78. https://doi.org/10.4000/medievales.503

FRC-IDF. 2017. *Valorisation de la venaison en Ile-de-France. Synthèse du projet à mi-parcours*. Fédération régionale des chasseurs d'Ile-de-France. www.ancgg.org/AD75/documents/rubd5_8147.pdf

Hell, Bertrand. 1994. *Le sang noir : Chasse et mythes du sauvage en Europe*. Paris: Flammarion.

Interbev. 2017. *Tous omnivores!* Interbev, Ipsos, Agence Vianova. www.interbev.fr/wp-content/uploads/2017/11/dp_omnivore_vdef-21112017.pdf

Jobard, Étienne, Jacques Marquay, Quentin Prigent, Stéphane Radureau, and Marie-Laetitia des Robert. 2016. *Évaluation du service écosystémique chasse en 2015. Rapport final de l'étude*. BIPE: Fondation François Sommer pour la chasse et la nature, Fédération nationale des chasseurs. http://docs.chasseurdefrance.com/FNC_BIPE_Rapport_Final_2016.pdf

Kimura, Aya Hirata. 2013. *Hidden Hunger: Gender and the Politics of Smarter Foods*. New York: Cornell University Press.

Larrère, Raphaël, and Catherine Larrère. 1997. "Le contrat domestique." *Le Courrier de l'environnement 30*. www.inra.fr/dpenv/larrec30.htm

Lavergne, Michel, and Emmanuel Faure. 2018. *GibSud de France. Caractérisation sociologique et économique de l'offre de gibier de chasse par les chasseurs dans le département de l'Hérault*. Montpellier: Fédération régionale des chasseurs d'Occitanie.

Lepiller, Olivier. 2012. *Critiques de l'alimentation industrielle et valorisations du naturel: Sociologie historique d'une "digestion" difficile (1968–2010)*. PhD thesis in Sociology, Université de Toulouse II-Le Mirail. http://agritrop.cirad.fr/583898/

Lepiller, Olivier, and Geneviève Cazes-Valette. 2018. "Supplementing Food for Health: Practices Amongst French Adults Aged 60 to 75 Years." *Review of Agricultural, Food and Environmental Studies* 99(3): 253-279. https://doi.org/10.1007/s41130-018-0080-y

Lepiller, Olivier, Estelle Fourat, and Élise Mognard. 2016. "Évolutions et facteurs socioculturels de la consommation d'aliments d'origine animale en France et en Europe: État des connaissances." In *Rôles, impacts et services issus des élevages et de leurs produits (Rapport d'expertise collective INRA-DEPE)*. Edited by Bertrand Dumont and Pierre Dupraz, 598-664. Paris: INRA.

Mouret, Sébastien. 2012. "La valeur morale d'un animal: Esquisse d'un tableau en forme de dons de vie et de mort. Le cas des activités d'élevage." *Revue du MAUSS 39*: 465-486.

Mouret, Sébastien. 2016. "Le véganisme." *Pour 231*(3): 101-107.

Poulain, Jean-Pierre. 2009. *Sociologie de l'obésité*. Paris: PUF.

Poulain, Jean-Pierre. 2017. *The Sociology of Food. Eating and the Place of Food in Society*. London: Bloomsbury Academic. www.bloomsbury.com/uk/the-sociology-of-food-9781472586230/

Scrinis, Gyorgy. 2008. "On the Ideology of Nutritionism." *Gastronomica 8*(1): 39-48. https://doi.org/10.1525/gfc.2008.8.1.39

SSP – FranceAgriMer. 2020. "La consommation de viande en France en 2019." *Agreste. Synthèses conjoncturelles 359*: 1-9.

Stépanoff, Charles. 2021. *L'animal et la mort. Chasse, modernité et crise du sauvage*. Paris: La Découverte.

Vialles, Noëlie. 1988. "La viande ou la bête." *Terrain 10*: 86-96. https://doi.org/10.4000/terrain.2932

Vialles, Noëlie. 1995. "De l'animal à la viande : Une mort sans cadaver." *French Cultural Studies* 6(18): 335-350. https://doi.org/10.1177/095715589500601805

Warde, Alan. 2016. *The Practice of Eating*. Cambridge: Polity Press. www.wiley.com/WileyCDA/WileyTitle/productCd-0745691714.html

2 From remorse to "hunter pride"

On a study of those who eat game[1]

Sergio Dalla Bernardina

The hunting world is changing at an astonishing pace. In just a few years, growing environmentalism has transformed not only the practices of hunters but also the memory of what they were. Gradually, the romantic image of the "enthusiast" drawn to the woods by an irrepressible urge has given way to that of the "sportsman," which in turn has been replaced by that of the environmental technician. Very recently, the "noble carnist"[2] image has also appeared, indicating that we may be seeing renewed pride among the minority of citizens who have not given up hunting.

Just as clichés have changed, so have practices and the ways in which they are represented. For researchers, this creates a risk of encountering historical reconstructions that, although perfectly sincere, anachronistically transpose the values of today's hunters onto the past. To complicate matters, the ecological context has also changed considerably, with the proliferation of ungulates everywhere, even in the regions of southern Europe where there was once a particular fondness for thrush pâté, ortolans marinated in Armagnac, or chaffinches with polenta.

In this respect, the sociologist's perspective is very valuable. It allows us to connect statements given by those who eat game with statistics, and thus helps us to better understand elements of continuity and change. Having worked on this subject, I read the study by Olivier Lepiller, Ophélie Roudelle and Sandrine Dury with great interest. Based on a clearly defined survey framework, the authors establish a picture of a particularly complex dietary dynamic. The extremely vivid accounts in their study inspired me to look again at the scenario I thought I had seen on the horizon around a decade ago. Today, here and there, its conjectural nature is apparent.

A highly hypothetical prediction

To contextualize this commentary, I will start by saying that in the preface to a collective work on the "call of the wild" (Dalla Bernardina 2012),[3] I made a daring

1 Translated by Cadenza Academic Translations.

2 Translator's note: In France, the image of the "bon viandard" has emerged. It recycles and reclaims a once pejorative term for an unscrupulous hunter interested only in killing as much game as possible.

3 Translator's note: Unless otherwise stated, all translations of cited foreign-language material in this commentary are our own.

DOI: 10.4324/9781003243496-5

prediction. Certain signs indicated that the emergence of "organic" culture and the familiarization of neo-rural groups with "spontaneous" food had given hunting a kind of second wind. I was encouraged in this interpretation by scattered data going beyond just the French context. Here is an extract in which I attempted to summarize my indicators:

> There is a vast category of users who do not hunt themselves (for moral reasons, because of the excessive cost of hunting licenses and the difficulty of the examination needed to obtain one, etc.), but benefit from the hunting done by others. Restaurants, supermarkets, and secondhand dealers are increasingly offering their customers "local" products coming directly from the woods: . . . The Swiss are eating more and more boar, red deer, and roe deer meat. Imports are booming, the shelves are flooded with farmed animals and frozen products. "Real hunting" enthusiasts denounce this industrialization. Hunting has arrived. It's impossible not to notice it: appealing slogans are filling butchers' windows, and supermarkets are making game their star seasonal promotion. From gastronomic restaurants to local pizzerias, game seems to feature on every menu. Once limited to high-end or mountain restaurants, it is now everywhere. It is a trend concentrated into a short period of the year, which for some is amplified by the desire for healthy, less calorific food: "Deer is fifteen times less fatty than lamb and boar is ten times less fatty than pig meat," explains Benoît Violier, hunter and chef for Philippe Rochat.[4] In Italy, in regions where the hunting tradition is more pronounced, game makes up around 10 percent of annual meat consumption.[5] This meat comes from hunting, but also increasingly from farming. In Liguria, an abattoir for wild animals has just opened.
>
> Although the French still seem a little hesitant compared to their Swiss or Italian neighbors (admittedly, the data are difficult to interpret, because game consumption does not necessarily go via the market), there have been several initiatives aiming to "normalize" this food (for example, in 2005, the laws restricting the transport and sale of venison were relaxed) and to change its image. One example is the creation of the label "Gibier de chasse, Chasseurs de France" ("Hunted game, Hunters of France"). According to the Fédération Nationale des Chasseurs (FNC) (French National Hunting Federation), this label is intended to highlight "the new possibilities available to professionals, with meat offering wild and forgotten flavors, both tasty and light, accessible to all." The FNC states that "With exceptional nutritional qualities, the ultimate natural product, rich in authentic flavors, game meat must find its place on France's restaurant menus again . . . and drive out a few

4 S. Gillieron. "Chasse: La saison de l'arnaque," *L'Hebdo*, October 7, 2009, www.hebdo.ch.

5 Anonymous, "La selvaggina trova nuovo spazio nella cucina italiana," *Catering News, Giornale della ristorazione e dei consumi fuori casa*, www.cateringnews.it/novita . . . /selvaggina-cucina-italiana/.

> misconceptions around game being a product that is strong-tasting, gamey, hard to digest, and slow to cook!"[6]
>
> (Dalla Bernardina 2012, 12–13)

Unfortunately, in my prediction, I had underestimated the increased (and growing) awareness of and concern for animal rights, and the intensity of anti-hunting sentiment in French public opinion. Ten years later, if we take the annual rate of game consumption by the French population as a parameter, it seems the "second wind" of hunting is not yet here. However, the situation is different in some countries outside of France. According to European statistics from the FAO, cited by Muriel Figuié and Raphael Malivel (2017), although the French and the Swedish significantly decreased their consumption of "other meats" between 1990 and 2013, the Germans and the British increased theirs.[7] Who is behind? Who is ahead of a trend? It's hard to say. When it comes to eating wild foods, it seems that disparate trends sit side by side. They coexist without necessarily converging.

A diverse picture

The study by Olivier Lepiller, Ophélie Roudelle and Sandrine Dury clearly illustrates the diversity of the motivations driving those who eat game and those who do not. They identify groups and rationales that vary depending on the proximity of those involved to the action of hunting: from the hunter/killer/cook, who shoulders the moral responsibility for killing, to the non-hunter who enjoys the taste of a good stag stew but prefers not to think too much about where it comes from.

The variety of statements collected confirms the idea that the domestic/wild opposition, this symbolic construction that has been with us since the Neolithic, is still effective today. It remains operational, but it is getting further from its concrete frame of reference: the rural world in which the "wild" was not yet something to be protected but a threat to be fended off.[8] It is in this revisited wild, a postmodern wild immune to the principle of non-contradiction, that the conceptual UFOs that have come into common use find their place, one example being the pair of terms "large hunted game/large non-hunted game."[9]

6 Press release by the Fédération Nationale des Chasseurs.

7 The French went from eating 6.6 to 3.7 kg/person/year, the Swedish from 3.1 to 1.93 kg/person/year, the Germans from 0.91 to 2.48 kg/person/year and the British from 0.2 to 1.54 kg/person/year. Clearly, the notion of "other meats" is rather opaque, but it does seem to refer mostly to meat from "wild" animals.

8 The domestic/wild opposition is consubstantial with the domesticating action of farmers in a permanent battle to protect crops (the wild being that which hinders the process of domestication).

9 Game is defined as "wild animals, birds, and fish that people hunt, usually for food." Therefore, no hunting means no game. Definition from Macmillan dictionary online, accessed 22 February 2022, www.macmillandictionary.com/dictionary/british/game_1#game_1__28.

Connections with the past

The analysis of the first group (producers/consumers) reminds us of the long-standing nature of the codes that governed the use and circulation of products from hunting in traditional society. Consumed by the hunter's family, game has always been given, bartered and eaten together at times of celebration in ways that follow the grammar of family relationships, alliances and friendships. Game circulated well beyond hunting circles. This is an important observation at a time when the connection between hunting and the rural world is sometimes questioned, even by the last representatives of the rural world, who now support the environmentalist cause.

Another theme that resonated with me was that of "gendered" orientations when it comes to food. Along with Alain Testart's famous study on women and hunting (1986), male preferences for "wild" meat reminded me of the statements I collected in Italy's Valle Germanasca in the 1980s. Here, I will quote the words of an innkeeper who welcomed the local hunting team on days when hunting was closed. His memories support the idea that the representation of hunters as predators and of hunting as a periodic return to the state of nature is far more than just a literary cliché:

> But chamois meat, generally, when you cook it, it's always jugged or . . . always in a stew, isn't it? . . . All its juices are in there, you eat polenta, that kind of thing. If Sandro is there, run for your life. If he has meat to cook, he roasts it. He doesn't want any juices. He wants it there, beautiful . . . roasted. But generally, almost everyone – *That is they prefer some juices* . . . – Almost, I don't know how to say it, but . . . really, there . . . au naturel, that's it, like, I don't know . . . like an animal eating another animal. Animals don't consider cooking it, they devour it there. And they're sort of like that when they eat. And they almost all eat it.
>
> (Dalla Bernardina 1994, 330)

Far from driving me to "substantialize" wildness (wildness being an inherent quality of certain beings and certain things), the results of this research led me to the following hypothesis:

> Are we to infer that hunters enjoy the key ingredient of their banquets for its "wildness"? This is what they claim, and perhaps even too emphatically, as if they were laying down a kind of rule: "Other people don't like it because it tastes strong. But that's exactly what we like about it." However, on closer examination, this loudly proclaimed preference is also revealed to be the result of a necessary choice, actually inspired by the family and its priority requirements. . . . [The "men's share" (Jolas 1982)] therefore seems to have been idealized and elevated retrospectively. The "wild taste" seems to have been reclaimed, transforming an initial handicap into an attraction, in the spirit of making a virtue out of necessity.
>
> (Dalla Bernardina 1994, 330–331)

Convergences around the relationship with identity

The study of the second group identified by Olivier Lepiller, Ophélie Roudelle and Sandrine Dury, whose members periodically buy game to reassert aspects of their identity and territorial connections, confirms the symbolic properties of the language of hunting, which is used well beyond hunting circles. On this, I found Mr. Castana's statement rather revealing:

> I associate boar with my village and my family. . . . It's not quite my Proust's madeleine, but it's not far off. . . . It's really associated with a cultural heritage, it brings back memories, it can be almost nostalgic.

In the introduction to the work *Manger moral, manger sauvage?* (Dalla Bernardina 2016), I proposed some thoughts that are entirely consistent with this, regarding the small wild animals that, unlike increasingly accessible large game meat, are no longer commercially (or legally) available:

> Among the many reactions, we can identify at least two "strong" ways of approaching the rise of "wild" values. On one side, there is moral objection (protection of the wilderness and giving up rare foods in the name of biodiversity). This is why species that, until recently, were enjoyed by gourmands and underpinned the gastronomic reputation of certain regions (for example the traditional dishes featuring thrushes, larks, chaffinches, and other small birds that were served in the Alpine Arc, from Gorizia to Sisteron) can no longer be hunted or sold. Today, eating an ortolan is just as criminal as smoking cannabis. On the other side, there is overconsumption (the incorporation of the symbolic and sociological properties connected to these non-reproducible resources). Those who eat finches, white-clawed crayfish, and elvers have not completely disappeared: they meet secretly, like conspirators, and indulge in their forbidden pleasures at clandestine feasts. Rural-dwellers sustain the tradition by giving their actions identity-related connotations.
>
> (Dalla Bernardina 2016, 4)

Hunters and flexitarians: a new alliance?

The discovery that, for some people, game is becoming palatable again because it is seen as "natural" was in line with my expectations. In the chapter by Olivier Lepiller, Ophélie Roudelle and Sandrine Dury, I was not surprised by the words of Mr. Géranine, who believes that "the meat's wild origin guarantees a degree of freedom from human intervention that can be harmful to health." In such cases, the interest in wild meat is not based on a concern for animal rights, but on health considerations.[10] I was also intrigued by the repeated references to the

10 This is reminiscent of those vegetarians who give up meat largely for health reasons.

"traditional" ethical model concerning (as among hunter-gatherers) never wasting game. I found traces of this myself in the statements of my sources, but combined with behaviors and sometimes words that were not necessarily compatible with it.[11] It is characteristic of the contemporary hunting world (as I stated right at the beginning) to mix real customs with clichés peddled by the specialized press, and even with discourses from visiting researchers.

What I had not imagined, and what is one of the key insights of the chapter on which I am building, was that the quasi-vegetarian minority of "flexitarians" would one day start eating game for moral reasons. Indeed, unlike livestock bred for meat, wild boar, red deer and roe deer have lived in total freedom. They can therefore be enjoyed without being complicit in any mistreatment suffered by livestock. This position is both consistent and unusual, and seems to give new arguments to the defenders of meat eating.[12] Hunting could benefit from this too, since it gives the practice an ethical respectability that its critics (at least those who eat animals that have lived in captivity)[13] would struggle to take away from it.

The animal's point of view

Following the rules of sociological tradition, the study by Olivier Lepiller, Ophélie Roudelle and Sandrine Dury gives an extremely neutral description of the dietary preferences and moral choices collected during the survey.

To conclude this commentary, just for the pleasure of intellectual provocation, I have allowed myself to break these rules and doubt the good faith of the social actors. I am not questioning the good faith of one category compared to another but *our* good faith in general. This is a working hypothesis. The comparative study of the cultural behaviors associated with "taking" animal or plant species reveals a kind of constant: taking a life invariably requires justification, be it mythical, ritual or based on customs, or whether ordinary discourse is left to justify it (Frazer 1983; De Martino 1975; Lanternari 1976; Burkert 1986; Dalla Bernardina 2020). The argument that presents a "good death" or "dying free" as the conclusion of a fulfilling life, appears, a priori, to be particularly moral (or at least this is my perception of it). However, it could be considered the latest variant of the countless ruses thought up by the human species to justify[14] irreversible acts that its conscience tells it require an explanation.

For a moment, let us put ourselves in the shoes of the wild animal.[15] It might say that today, protected by the law, fed when it snows, monitored by cameras and tracked using GPS technology, it is no longer wild at all. It might ask us to leave

11 I am not questioning the existence of this norm. I simply believe that not all "traditional" hunters act in accordance with it.

12 New but particularly restrictive arguments.

13 According to the survey by Cazes-Valette (2007), 40% of those who frequently eat game say they are against hunting.

14 Psychoanalysis would call this "rationalization."

15 I know this is just a rhetorical expedient, but everyone speaks for animals these days.

it in peace, emphasizing that we have no authority to decide what is good for it. It might add that this idea of a happy death rounding off a life in the great outdoors is an old promotional expedient and cite the following slogan: "There's something truly amazing about Auchan life. Before it is caught and sliced, Captain Sea smoked salmon from Brittany has swum 30,000 kilometers against the current. That shows it has made the most of life!"[16] (Dalla Bernardina 2020, 106–107). However, for as long as we (like the author of this short commentary) remain attached to the meat-eating diet, we will find it hard to listen.

References

Burkert, Walter. 1986. *Homo Necans: The Anthropology of Ancient Greek Sacrificial Ritual and Myth*. Translated by Peter Bing. Berkeley, CA: University of California Press.

Cazes-Valette, Geneviève. 2007. *Le rapport à la viande chez le mangeur français contemporain. Rapport de synthèse*. Toulouse: Groupe ESC/CCIT.

Dalla Bernardina, Sergio. 1994. "La part des hommes ou les rebuts de la famille? Les emplois du gibier dans le Val Germanasca." In *Gens du Val Germanasca. Ethnologie d'une vallée vaudoise*. Edited by Sergio Dalla Bernardina and Danièle Dossetto, 319–346. Grenoble: Centre alpin et rhodanien d'ethnologie (CARE).

Dalla Bernardina, Sergio, ed. 2012. *L'appel du sauvage. Refaire le monde dans les bois*. Rennes: Presses Universitaires de Rennes.

Dalla Bernardina, Sergio. 2016. "Introduction. Le mangeur et le bon sauvage." In *Manger moral, manger sauvage?* Edited by Sergio Dalla Bernardina, 3–17. Paris: Comité des Travaux Historiques et Scientifiques.

Dalla Bernardina, Sergio. 2020. *La langue des bois. L'appropriation de la nature entre remords et mauvaise foi*. Paris: Muséum National d'Histoire Naturelle.

De Martino, Ernesto. 1975 [1958]. *Morte e pianto rituale. Dal lamento funebre antico al pianto di Maria*. Turin: Bollati Boringhieri.

Figuié, Muriel, and Raphael Malivel. 2017. *La consommation de viande de chasse en France et en Europe. Synthèse bibliographique*. Montpellier: CIRAD-ES-UMR MOISA.

Frazer, James George. 1983 [1923]. *The Golden Bough: A Study in Magic and Realism*. London: Macmillan and Co.

Jolas, Tina. 1982. "La part des hommes. Une société de chasse au bois." *Études Rurales* 87–88: 345–356.

Lanternari, Vittorio. 1976. *La Grande Festa. Vita rituale e sistemi di produzione nelle società tradizionali*. Bari: Edizioni Dedalo.

Testart, Alain. 1986. "La femme et la chasse." *La Recherche 17*(181): 1194–1201.

16 Advertisement from the mid-1990s by Auchan (a supermarket with a particularly strong presence in France).

Part 2

Promises and market

Conversation 2. Producing natural wines

3 "Natural" wines

The call of the wild grape[1]

Christelle Pineau

Introduction

Inverted commas or no inverted commas? We must ask ourselves this question. Just like the word *nature* (Descola 2013 and 2018), the word *wild* is problematic in terms of its usage. Overused, with too many connotations, overflowing with different meanings depending on the place of its use and on the period associated with it, *wild* has become an everyday word, common in the contemporary spaces of our daily lives. It is exploited and given positive connotations in marketing (to advertise a major fragrance brand).[2] Conversely, it is used with negative connotations in political debates (to evoke forms of incivility, precarity and hubris).[3] Whatever the chosen connotation, *wild* shows the intrinsic power of words, just as the *wilderness* populates our imaginaries today, like a logical or necessary counterpoint to our post-rural societies (Dalla Bernardina 2011). The concept of *wilderness* in the English language comes out of a certain vision of the untouched natural space that is connected to the colonization of America (Larrère and Larrère 2015).[4] It is just as richly and diversely evocative as its French cousin *sauvage* (meaning "wild"). Both concepts are pursued in contemporary art, captured on screen and on paper, and summoned by city-dwellers who are cutoff from the living world. Just as the romanticization of *nature* is "very widespread"[5] (Tsing 2020, 250), the romanticization of the *wild* and *wilderness* also occurs. However, as Anna Lowenhaupt Tsing states, "understanding its global spread does not mean naturalizing it by making it a universal biological trait" (back-translated from the French).

1 Translated by Cadenza Academic Translations.

2 The luxury brand Dior made American actor Johnny Depp the face of its fragrance "Sauvage," combining the imaginaries of the wilderness and that of the "wild" by choosing to keep the name (which means both "wild" and "savage") in French. An advertisement for this fragrance sparked controversy when it was accused of cultural appropriation for its Native American imagery. This case indicates that the perception of "wild" or of "wilderness" swings between positive and negative.

3 Particularly in the French political context of 2021, preceding the 2022 presidential election.

4 On this point, see the chapter "Sauver le sauvage? L'idée de *wilderness*" (Larrère and Larrère 2015, 25–50).

5 Translator's note: Back-translated from the French.

DOI: 10.4324/9781003243496-8

Taken as subjects of reflection, *nature* and the *wild* will be examined here to help us understand the presence of the "wild" within "natural" wine production. To answer the initial question, I will henceforth be putting the word in inverted commas, because of its great variability and to avoid any oversimplification (be it positive or negative). It is also important here to differentiate between *wilderness* (a space) and the "wild" or "wildness" (terms referring in this analysis to that which is wild or to the wild state).[6] The "natural" wine movement took off in the 1980s and intensified in the 2000s to form what today has become a real international network. Winegrowers make wines that are described as "natural" (another word that has come into the civil sphere and is used by consumers and part of the sector). The presence of "natural" wines in France and worldwide remains inversely proportional to their renown.[7] In preliminary works (Pineau 2017, 2019, 2021), I attempted to show the meaning of these approaches that use vineyards and wine cellars. Without synthetic chemicals or excessive manipulation of the must, they bring together different types of knowledge, from the most academic (e.g., using a microscope to observe fermentation)[8] to the most empirical, concrete and spiritual (using medicinal plants, employing the lunar or biodynamic calendar or borrowing from forms of shamanism to communicate with the vats). This text aims to examine these practices through the lens of the "wild" and to reveal the various forms in which it is present. It could have been nothing more than a stylistic exercise or a rhetorical game, were it not for around a decade of fieldwork in the French vineyards that produce "natural" wines. This fieldwork provided more concrete elements to support this reflection: first, the plant itself, in its "ethos" and its millennia-old relationships with humans, seems

6 Translator's note: In the original French version of this chapter, the author used the English word *wilderness* but chose to preserve its specific genealogy by keeping *sauvage* (meaning *wild* or *savage*) in French. For ease of understanding, the choice has been made here to translate *sauvage* as *wild*.

7 "Natural" wines go further than organic agriculture. They currently have no legal status. Nevertheless, a private charter has been drawn up by the Syndicat de défense des vins natur'L (Union for the Defense of Natural Wines). Created in 2019, it outlines the principles of the practice. This charter has been assessed by France's Direction générale de la concurrence, de la consommation et de la répression des fraudes (DGCCRF) (Directorate General for Competition Policy, Consumer affairs, and Fraud Control). The DGCCRF proposed amendments to bring it into compliance with its rules. There are no statistics on the number of "natural" winegrowers, because of the many biases, particularly relating to the difficulty defining and delimiting "natural" wine and to the way in which the winegrowers describe themselves. For example, some winegrowers can be "natural" and biodynamic, meaning that they only have the Demeter or Biodyvin label, or they may have no labels and not be part of any associations. Nevertheless, a few data suggest a figure of around 1%, insofar as the global production of organic wines is estimated at 3–5% (source www.agencebio.org). "Natural" wines can also be associated with biodynamic agriculture, a holistic agronomy principle theorized by Rudolf Steiner (1861–1925), founder of anthroposophy. The anthroposophical movement is still subject to controversy, particularly because of its esoteric approaches.

8 Jules Chauvet (1907–1989) – tutelary figure of "natural" wines, winegrower and wine merchant and a researcher in microbiology and chemistry – tested wines without added sulfur from 1951. In particular, he developed the use of the microscope to understand the different fermentations. Paul Bréchot, former fermentation director at the Institut Pasteur, was his research partner.

to retain a "wild" force. Second, the reality of this type of wine production links to evocations of the "wild" in its harshness, its difficulties and its fragility. Two sides emerge: one more exalted with positive connotations, the other harsher with negative connotations. We will see how the "wild" permeates "natural" approaches, like a palimpsest, and then, we will ask whether the old dialectics of the wild and the domestic survives these practices. Could the lines between these categories be starting to blur? Could we be witnessing a reverse evolution of one toward the other? Or, to use the words of Tsing regarding the forests of Borneo and the Meratus Dayak, could we be heading into a "gap between cultivated and wild," (Tsing 2020, 288) in other words, a gap in which hybridization may never have stopped?

1. A paradoxical plant

Grapevines are polymorphic. There are two sides to them: in their wild form, they are creepers, but over millennia of cultivation, human actions have changed their physiognomy. Their two "ontologies" form two layers, reminding us of the homeostatic nature of a plant with a paradoxical identity.

1.1. Daughter of the woods

From the terrace of my garden, I observe the vine leaves perched ten or so meters above my head, sometimes next to oak leaves, sometimes Douglas fir needles. This image reminds me that grapevines are creepers, epiphytic plants that climb along trees and curl around other alter-organisms that support them. I don't know where this vine comes from or what species it is. Did it escape from an old, abandoned, neighboring vineyard? If so, could we say it has "rewilded" itself? This reflection takes me back to my land in the Coteaux d'Ancenis (Loire), to the time when a young winegrower said to me: "An old man told me there are wild grapes along the Loire. I walked in the area the other day, looking for them." The excitement in her voice implied that wild grapes offered something extra: greater authenticity, or more flavor, perhaps? Or maybe the freedom of a vine that has made its escape, claiming back its independence? At the time, I wasn't sure, but I observed this sense of wonder again very recently, at a "natural" wine trade fair in the Luberon. I heard the message several times that day, from both professionals and amateur tasters: "Have you tasted the wild grape wine?" Clearly, the appeal of the "wild" was at work. And no, I didn't get a chance to taste it. The wine being rare, there was none left when I finally got near the coveted stand. Would I have judged it as objectively as possible in this context? Or would I have succumbed to the "wild utopia," like many urbanites and semi-urbanites, thereby contributing to the intensification of the phenomenon (Dalgalarrondo and Fournier 2020)? It's impossible to say, but I remember this repeated positive evocation, an increased appetite for the "wild" in the "natural" community. It was reminiscent of the late-summer gleaners who engage in wild harvesting, picking grapes on abandoned plots to make "clandestine natural" wines for their own consumption.[9] In a sense, it corresponds to the emancipatory journey away from the dominant norms of modern, productivist winegrowing and winemaking, with some winegrowers rejecting these norms and choosing "natural" practices. Thus, the winegrower is intertwined with the vine, united in the reclamation of the self and of their choices. This escape allows them to recover a kind of creative energy (Gorz 1988), far from certain normative frameworks prescribed in particular by the appellations d'origine contrôlée (AOCs) (Controlled Designations of Origin).[10] Many of them experience the system of controlled designations as a kind of straitjacket, in that wines are oriented toward a taste typicity specific to each appellation, most often with the help of enological products that these winegrowers reject. This is expressed by the words of an Ardèche winegrower remembering his conversion from "conventional" to organic then "natural" wines:

> I was trapped, almost imprisoned. . . . Every year, I work as if it is my first vinification, as if I have never made wine before. I have the freedom to do this. I don't have to think about what people will or won't like.[11]

The tendency to associate "wild" and freedom is often observed in the world of "natural" wines. The same goes for the character of the vine-creeper that escapes and tries to reclaim its autonomy, and for the idea expressed by this winegrower from Anjou:

> It's an elusive plant. Even if you try to stop it at some point by some idea, be it pruning, wine, working the soil, in fact it's an elusive plant, it will always escape us, and that's a good thing.[12]

9 This is not a main occupation for these gleaners. Others, more structured into associations such as Marseille's C.U.V.E. collective, may combine gleaning and grape buying, to sell some of their produce (e.g., at cultural events). On this, see the article published in March 2021, in *L'empaillé* (a newspaper in France's Occitanie region): Des Chourragnans, "Fragments d'histoires de vin clandestin," photo: Lise.

10 Appellations d'Origine Contrôlée (AOCs) (Controlled Designations of Origin) were created in 1935 in France, in a context of overproduction, poor wine sales and fraudulent practices. Today, there are 357 AOCs for wine. Seen as guarantees of quality when they were created, they are criticized by many "natural" and/or biodynamic winegrowers, who think that they have been distorted. In 2001, in France, this led to the creation of the group Return to Terroir, which has members from 13 different countries. Source: "Biodynamics, viticulture for tomorrow," Return to Terroir, accessed January 28, 2022. Available at: https://renaissance-des-appellations.com/en/.

11 Gilles Azzoni, extract from the documentary "Anthropologie des vins 'nature,' la réhabilitation du sensible," (Pineau 2017). Translator's note: Unless otherwise stated, all translations of cited foreign-language material in this chapter are our own.

12 Didier Chaffardon in "Anthropologie des vins 'nature,' la réhabilitation du sensible," (Pineau 2017). Director of the documentary.

Nevertheless, all these winegrowers remain aware of the paradoxical situation of their vines. Despite being cultivated using natural methods, they are the products of clonal selection, and the vast majority of them are coupled with American rootstocks. This situation is the legacy of the phylloxera crisis in the late nineteenth century. Vines that are "own-rooted" (ungrafted because they were not destroyed by the damaging American pest) are in the minority, despite propagation from cuttings here and there. The experiments are not always successful, and the risk of disease is always present. What's more, they are unauthorized and are kept discreet. The idea of "wild" grapevines evoked earlier refers more to vines that have "escaped" from their plots. In fact, it is precisely this ability to recover an original form (the creeper that grows along other plants) that enthralls all winegrowers, a fortiori all "natural" winegrowers, who have entered into a relationship of relative mastery over their vineyard. They are struck by the contrast between the supple creeper and the hard-trunked, bonsai-like little tree into which it has been transformed, leading them to exalt the wild form. Similarly, the sight of old vines that spontaneously propagate by layering,[13] without human intervention, fuels the rhetoric of emancipation and autonomy. However, these projections, which sometimes fall within the realm of metaphorical analogy, are in fact rooted within the genealogy of the plant. The wild grapevine is not a fantasy. It does exist. Its name, *vitis sylvestris* (or wild grape), explicitly evokes its origins as a creeper sheltering in forest environments. It interests winegrowers both for the flavor of its wines and for its resistance to disease. This resistance is confirmed by various research teams working on climate change and biodiversity decline, including a Franco-Swiss team. Beyond the observation that European wild grapes are becoming rarer, this plant is useful from a genetic point of view:

> Currently, the conservation of wild forms that are close to cultivated plants is indispensable for maintaining genetic variability and combating genetic erosion. Vine monoculture covers vast areas, which also increases the risk of epidemics. The genetic characteristics of wild grapes can be essential if this occurs. They have long been known to be resistant to frost.[14]
>
> (Arnold, Gillet, and Gobat 1998, 159)

This rarity makes wild grapes even more valuable in the eyes of "natural" winegrowers, who are the frontline observers of declining biodiversity and are committed

13 Propagation by the mother plant. In the case of vines, a branch can independently root into the ground and grow a new main stem a little further along. This operation can also be encouraged by the winegrower.

14 See also the website of the Institut français de la vigne et du vin Occitanie (the Occitanie French Wine and Vine Institute), which confirms the increasing scarcity of wild grapes and the interest in research: www.vignevin-occitanie.com/fiches-pratiques/les-lambrusques/, accessed 28 January 2022. The Marseillan Institut national de recherche pour l'agriculture, l'alimentation et l'environnement (INRAE) (National Research Institute for Agriculture, Food, and Environment) has a botanical conservatory of around 400 specimens.

participants in its preservation. The quest for *vitis vinifera silvestris* among "natural" winegrowers takes on another dimension here, which is more pragmatic than symbolic. Beyond mythical tales of the original pure wine, etiological data (in the strict medical sense of the term referring to the origins of diseases) give this notion of "wild" increased interest. Already, this dual-focus perspective goes some way toward reducing the distance between the wild and domestic categories.

1.2. Intertwined with the human

Grapevines and humans are an old pair. The incarnation of the plant is reflected in the anthropomorphic images used to describe vines in the French language: Vines have eyes (*yeux*: buds), arms (*bras*: branches) and a foot (*pied de vigne*: vinestock). Their shoes are put on (*chausser*: hilling up soil around the base of the vine in fall and winter to protect it) and taken off (*déchausser*: the opposite process in spring). Thus, vines intertwine with the human body, "demonstrating our symbolic relationship with the vine" (Amiel, Charuty, and Fabre-Vassas 1981, 41). The words reveal the constant proximity constructed between plant and human since they first came into contact. Although the origins of the first wines remain unclear, we know that the grape–human pair preceded them:

> There have been grapes, and people to gather them, for more than two million years. It would be strange if the accident of wine never happened to primitive nomadic man. But before the last Ice Age there were people whose minds were far from primitive. Such high intelligence, such organization and aesthetic sense as the Cro-Magnon people had to paint the masterpieces of the Lascaux caves, in French forests where the vine still grows wild – although now an escape from vineyards – suggests that wine could have been known to them, even if we have no evidence one way or the other.
>
> (Johnson 2020, 40)

In fact, "the oldest pips of cultivated vines so far discovered . . . were found in Soviet [as it then was] Georgia, and belong to the period 7000–5000bc" (Johnson 2020, 40). Now, several millennia later, in the pruning season in January, a winegrower near Angers explains his choices. He says that he prefers goblet pruning to trellis pruning,[15] because it feels more alive to him:

> Goblet pruning is very good for air circulation. There's never bad rot because it's well ventilated, and when it's pruned, this is what you get . . . *[he shows the remaining cut wood, which forms a hand]* Obviously, when I see that, I don't know, it . . . (laughs) I get kind of emotional! You see little hands like that emerging. It's a big moment! . . . Every winegrower should go and prune a

15 Goblet pruning favors several branches on either side of the vine. Trellis pruning favors one main branch that is then curled around a wire.

> hectare, if not more, but they absolutely must go and do it. It's a connection you lose if you don't. . . . That's the aim: you try, from something shapeless, to make a kind of hand, taking into account the currents of sap to avoid damaging the vine.[16]

In this type of natural practice, reference is made to the imagery of the hunter-gatherer, either knowingly or unknowingly, depending on who is speaking. By getting as close as possible to the vines (not using harvesting or pruning machinery), the "primary" and simple act of picking is linked to a past that, although real, is also mythologized. The action becomes a tool for synchronicity and strengthens the idea of permanence in the relationship between vine and Human. Thus, beyond the winegrower's objective (which is of course to harvest grapes and make wine), affective relationships are formed with the plant, and these can be intense. As we have already seen, the words used to embody and even anthropomorphize the plant once again emphasize the great proximity between the two living "kingdoms." This way of expressing the relationship to vines is far from an isolated case. It is even commonplace among these winegrowers. These sentimental stories are only told after significant discussion, not only because they are personal but because of fear that they will be poorly understood or that the person hearing them will not share the sentiment. Another winegrower in Pyrénées-Orientales, also part of the "natural" movement, became similarly emotional when explaining his pruning method. We spoke in March, in the heights of the Banyuls wine region:

> We regulate the load on the vine, we help it to live, because if we ask too much of it, it won't do well, and if the load is unbalanced in the other direction, it will be too vigorous, it will be out of control and make shoots all over. Pruning helps you understand the vine, whether it's doing well, whether it's not doing well. That's why, when I hear people talking about harsh pruning, for me, it's mind-boggling! If you're doing that, you're seeing yourself as a producer, it's like you don't think you're working with a living thing. It's a plant but it's alive, it's living! It has a life, a real life. It's like animals in factory farms. It's the negation of life. I do this job precisely because it's alive. I work with living things. Wine is alive, fermentation is an extraordinary thing, it's a transformation with yeast, with microorganisms. Soon, the vine will be flowering, the buds will come out. To start with, you see clusters of flowers, which will blossom and become grapes. . . . It's poetry! So if you've lost that relationship with the plant it's *[swears and laughs]*!![17]

The exaltation of the living seen here is a semi-conscious appeal (the level of awareness varying between winegrowers) for the original form of the vine: wild

16 Extract from an interview in the vineyard with Mark Angeli, January 2015.
17 Extract from an interview in the vineyard with Alain Castex, March 2014.

grapes. All the discourses produced on the wild grapevine's ability to protect itself, to grow, to develop independently and to become autonomous refer to an intrinsic power believed to be stronger because it has not been distorted by domestication. For these people, as for Gaston Roupnel in his *Histoire de la campagne française* (written in the first half of the twentieth century), the plant *vitis* contains, beyond its domestication, the notion of wildness:

> The whole appearance of the vineyard betrays this independence, this creativity, this turbulence of the establishment. Only untamed spaces are ruled by vines. Overcoming the resistance and ingratitude of the soil, vines seem to slip and creep in everywhere, taking root, governed only by the whims of their irritated tendrils. They appear everywhere, winding their way between stones, happily free from pruning, following the chaos and caprices of their laughing, twisting foliage.
>
> (Roupnel 1974 [1932], 196)

The lyrical words of historian Gaston Roupnel, a writer and a winegrower, express the "wild" ontology of *vitis* when it is given the right conditions. A few decades later, on the limestone plateaus of Languedoc, a winegrower who is viscerally attached to "the things of nature" evokes and promotes the "wild" nature of her vines, which she cannot bear to take away from them:

> I see them as quite wild, which means I try to intervene as little as possible, and when I do, I always question myself, you see. I think "what, am I going to meddle to do that?" It's above all pruning that bothers me. Deep in my heart, I've still not resolved the conflict. . . . When you prune, you do it to encourage fruit, and to keep the vine alive too, because if you never prune it, it eventually dies, it makes wood, wood, and more wood, and then there are no grapes. Because for cultivated vines, it's humans who have made them like that, and now they're dependent on it, and to continue living with fruits and flowers, they need pruning, and I really struggle with that. And I think about lots of things, about the kind of pruning, much gentler, leaving more wood, but the vines don't necessarily lend themselves to it. In fact, I don't know, I've not found a solution yet. My aim is for the vines to be totally autonomous, independent, and wild, as I like to be myself![18]

The intertwining of the self and the vine is expressed plainly, as if the vine were an extension of the self. The vine becomes a mirror of the self, and more broadly a reflection of the world that seeks to return to a "wild" state, "wild" having positive connotations here, associated with autonomy and independence, and therefore with a non-hierarchical view of the world. The possibility of wildness does not frighten this winegrower, who, as far as possible, rejects the taming of the plant.

18 Extract from an interview with Anne-Marie Lavaysse, February 2014.

In this way, she positively reevaluates "wildness," seeing the historical evolution toward domestication as a restraint of the living. This interpretation contrasts with dominant Western thought, as Sergio Dalla Bernardina explains:

> Among Western farmers, . . . the wild state is viewed as a "before" and in this sense as a negative reality, in that this "before" is trying to catch up with and replace the present, hindering the process of domestication. But it is also seen as the place of the origins, of potential that is waiting to be refined, the place of the ancestors and the founding heroes. Its presence, if only on the logical level, is indispensable for cosmic balance.
>
> (Dalla Bernardina 2011, 23)

For this Languedoc winegrower, the idea of wanting to rely on what vines are capable of giving by themselves amounts to accepting a form of determinism and subjection in the relationship to "nature," while rejecting subordination to the productivist model in which plants (and humans) are enslaved. By this logic, the dominated–dominant relationship becomes a moral contract: by not excessively exploiting vines, by limiting our interventions to "gentle" care, humans can in return harvest the fruits of this collaboration without guilt. The relationships of dominance are therefore modified: by agreeing to give up some of their control and adopting the principle of supporting vines, and the wine produced from them, as living things, winegrowers find themselves partly in the role of the dominated. Thus, the winegrower and the vine each leave the other some freedom of action in their relationship.

2. The presences of the "wild"

Although certain winegrowers sometimes employ the word "wild" in conversation, it is not held up as the banner of "natural" wines.[19] Jules Chauvet (1907–1989), a winegrower and wine merchant in Beaujolais, and a researcher in microbiology and chemistry who became a tutelary figure of the "natural" wines movement, did not use this word to describe wines without synthetic chemicals or added sulfur. Neither did he use the word "natural," preferring to describe this kind of wine using the criteria "purity, terroir, harmony" (Néauport 1997). In this case, why push the "wild" characteristic in "natural" wine production? Once again, my fieldwork indicates that this notion may reflect the harshness and uncertainty of the practice. This section will therefore examine the harshest dimensions of "natural" wine production, with reference to the most somber side of the "wild" corpus.

19 Nevertheless, winegrower Sybil Baldassarre, in the Faugères region (in the southwest of France), has chosen to call her estate "La Graine sauvage" (The Wild Seed).

2.1. A tough model

For some, going into "natural" wines can be experienced as an irresistible call for free expression of wine and of the self. For others, it may be the final step in a progressive conversion via several stages (from conventional[20] wine production to organic agriculture, then to the "natural" approach, which goes further than organic principles). It can also be a combination of the two. In all cases, it is a difficult and risky move, both physically for winegrowers and financially for their small businesses (Pineau 2021). Growing vines without synthetic chemicals (pesticides, fungicides, etc.) is a gamble, with the odds changing every year depending on the weather conditions and the acceleration of climate change. A winegrower in Touraine tells of her early days in natural wines and the frugality of her daily life:

> In fact, this is the first year I've had all my wines. Let's just say I'm managing to pay back my loans, pay my suppliers, and run the shop. . . . It's working, just about, but it's very precarious. Still, apparently, it's like that for everyone. . . . In any case, I don't need much. I'm not interested in clothes, my truck is my wheels and it's fine as long as it moves, I don't. . . . Actually, yes, I need vegetables![21]

Most winegrowers work on a few hectares, often alone (working more than three or four hectares alone becomes very difficult). Because issues cannot be rectified with enological products, optimal harvest quality is essential, as is strict sorting if the grapes are affected by disease. This can reduce the yield, which means fewer bottles to sell. It becomes even more difficult when newly established winegrowers suffer a string of bad weather. Without a stable cash flow, the "natural" venture can quickly collapse. Making "natural" wine is like "working without a safety net," in the words of Gilles Azzoni, a winegrower in Ardèche. Furthermore, this practice requires more time to be invested in the vines (50% more, according to some estimates) than in conventional viticulture. The time spent working the land, grinding compost, and spreading biodynamic preparations[22] (for those who combine biodynamics with the "natural" approach) or other types of preparations based on plants or essential oils considerably increases the number of hours spent working in the vineyards. It is also physically arduous. The body, "man's first and most natural instrument" (Mauss 2013 [1936], 372), reinstated in the face of the

20 Certain "conventional" winegrowers prefer to describe their wines as "classical" or "traditional." They reject the term "conventional," which has negative connotations compared to organic agriculture. Personally, I will continue to use the word "conventional," because these practices are the result of conventions and because the term avoids confusion. The terms "classical" and "traditional" are too vague and could just as easily apply to organic or "natural" wines.

21 Extract from an interview with Marie Thibault, July 2014.

22 This can be the spreading of the best-known preparation: "BD500" or "horn manure." This is made by packing cow manure into horns, then burying these over winter to make humus. This preparation is then stirred in alternate directions for an hour, before being sprayed in tiny quantities on the vineyards to improve the soil.

machine, must become harder and tougher. Many recall their first months facing the challenges whatever the season and building physical endurance. Counter to an agricultural model that promotes ever bigger and more technologically sophisticated machines with little connection to the land, manual work (sometimes combined with the presence of animals such as sheep, pigs and geese, or with the use of horses) is rehabilitated, in all its difficulty, stripped back to the bare essentials,[23] but also in all its pleasure, because it allows direct contact with the plants. The proximity, the time spent observing and the hours literally pouring sweat into the work all allow a closer analysis of the signals given by the vines. The senses are awakened and tested, particularly through touch and smell, which are often undervalued in modern society, dismissed as animal and "wild."

The winegrowers can also be associated with the "wild" category in the sense that they are disparaged, marginalized or even denied by a section of the wine industry that dislikes these troublemakers who threaten the dominant model. The system of designations of origin, of which the initial objective[24] has been distorted according to the "natural" movement, is being shaken right to its foundations, almost ordered to justify its norms, which are judged to be obsolete in the contemporary context of climate change. Occasionally, these tensions become legal battles, and the hearings and media coverage shine a spotlight on the huge conflict between the two communities. "Natural" winegrowers are very much a minority, but the community is growing and building itself a reputation among consumers, while the upholders of conventional viticulture and AOCs are in the majority and maintaining their place, but feel threatened. The two sides sometimes engage in discussions, but some more belligerent players in the conventional sector bring "natural" winegrowers before the courts. Several emblematic trials have illustrated this frontal opposition.[25] The stakes are high, given that even a partial challenge to the dominant model has an impact on various levels of the sector, from the practices (which affect companies making phytosanitary products, equipment manufacturers, etc.) to the institutions (AOC, viticultural syndicates, customs, professional competitive examinations, etc.), and to teaching. Winegrowing and winemaking schools are still built upon the conventional model, even though some establishments and teachers are (to varying degrees) becoming more open. Organic agriculture, alternative treatments and even biodynamic agriculture are evoked but mostly in the form of modules. A young winegrower told me that around 15 years ago, in the teaching room of his viticultural school in Narbonne, there were posters advertising Monsanto treatment products, which have since become highly controversial and even been banned in some countries. Already convinced by organic and biodynamic agriculture, he remembers the spirit of a

23 "Natural" winegrowers may also have efficient tractors for certain operations.

24 French AOCs were created in 1935 to combat fraudulent practices.

25 One example is the case of Olivier Cousin in Anjou, who was brought before the courts by the Anjou and Saumur winegrowers' federation in 2014 for using the word "Anjou" on his labels, which is unauthorized for wines that do not qualify for the AOC. He was ordered to pay a symbolic euro to the plaintiffs after an appeal hearing by the Court of Angers in 2015.

slogan that struck him: "Treating eliminates better." The elimination of "weeds," plants that grow in a disordered or even "anarchical" way, comes out of a kind of prophylaxis linked to monoculture and to a kind of rigid thinking that defines the conventional model and pesticide use as viticultural norms. Any other practice that does not use these chemicals is again dismissed as wayward and rebellious thinking. Yet weeds, the target of the most toxic products in modern agriculture, can play a role in the ecosystem of vines. "Natural" winegrowers have various approaches to them, depending on the context, sometimes eliminating them with labor techniques (tractor or horse) if they create too much competition for struggling vines, sometimes leaving them and rolling them to provide mulch in the row and to encourage biodiversity. Moreover, there are different types of weeds, the most troublesome (dominant) being those that have become resistant to phytosanitary products. This is observed by winegrowers when they are converting plots that have been treated with synthetic chemical weedkillers to use them for organic agriculture. Therefore, there are multiple kinds of weeds, which become entangled and can be "wild," in that they do not fit a monolithic vision of viticulture where yields take priority. The "natural" method, seen as "wild" by those who do not use it, is itself the product of "wild thought" that experiments and chooses freely.[26] Everyone has this capacity for "wild thought" in them, although they may express or repress it to different degrees, as Lévi-Strauss reminds us in the work that explicitly echoes our reflection: "Wild thought is not the thought of wild men, but the wild power of all thought, insofar as it has not been domesticated for its yield" (Lévi-Strauss 1990 [1962], 262). This ability to bring out the "wild" part present in every human mind is embodied throughout the entire production chain of "natural" wines: at the start of the project, in the vineyards, in the experiments and in the glass.

2.2. *Smell, drink and read*

The materiality of the "wild" can sometimes be apparent in the glass if animal notes are present when the wine is poured. The grape skin is home to a whole yeast population that is useful for natural fermentation (meaning that indigenous yeast is used, with no exogenous additions). A certain yeast can also be present and can give the wine a certain animality if there is too much of it (e.g., Brettanomyces). This can give the wine "odors of the stable," which opponents of "natural" wines are keen to criticize. This stigmatized animality, a symbol of uncontrolled

26 A retired conventional winegrower from the Anjou region, who could not bear the sight of weeds in vineyards and the way these vineyards were run, told me on the subject of a "natural" winegrower colleague: "In his place, I would have pulled up all the vines and replanted!" He saw the vines as neglected and accused the winegrower of being lazy. His idea of "clean" vines (maintained via the use of chemicals) corresponded to his model of tamed vines that stayed in line. Words taken from an informal discussion in the vineyard, June 2021.

vinification, is repellent to lovers of domesticated or "civilized"[27] wines. These odors are not always present, and when they are, they fade or disappear after a variable decanting time. The animal odors can be a result of the phenomenon of reduction, or of autolysis of the indigenous yeasts, which means that the wine needs aerating. These occasional odors, which depend on the batch, the year, the bottle, the estate, etc., fuel the controversy among conventional wine tasters. Yet the language of animality, in its most "wild" aspect, exists in the interpretive framework of conventional tasting. There are notes that evoke game, hare, civet, leather or musk.[28] These animal notes are in fact valued as the expression of a certain quality of the wines, particularly "old Burgundy or certain Bordeaux or Bandol wines," according to the *dico du vin*[29] website. This demonstrates that the rhetoric of animality can be used both to favorably describe and to harshly discredit, depending on the intention of the person using it. When it is used in conventional wine circles to target "natural" wines, it often aims to discredit a practice underpinned by a worldview that seeks to counter complete control of living things. Meanwhile, the world of "natural" wines accepts these references to animality without pejorative connotations. We can therefore observe that the rhetoric of animality, embedded in that of the "wild," is neither stable nor neutral. Its connotations can swing from negative to positive, as shown by the works of Max Léglise (1924–1996), a Burgundy-based enologist and engineer at the Institut national de la recherche agronomique (INRA) (French National Institute of Agronomic Research) in Beaune from 1948. In the course of his career, he focused on new vinification methods in agrobiology, and came to the conclusion that vines "on the verge of their wild state" gave better results in terms of taste (Léglise 2007 [1990], 16–17).[30]

Are "wild" characteristics in wine a fault, a deviance or a sign that the wine is alive? Although we sometimes see temporary reconciliations over a bottle between conventional wine tasters and "natural" wine tasters, the issue remains sensitive. Given that it engages people from an ontological perspective, it may even be unsolvable. These fundamental distinctions are particularly evident on wine labels. This can be seen by simply stepping inside a specialist "natural" wine store. "Natural" winegrowers express themselves on these few square centimeters of paper, showing their choices, their difficulties and their dispositions. They assert their distinction through humor, poetry, wordplay, and provocative or political messages. One example is Jean-François Chéné's "Les Joues Rouges,"[31] which evokes the Anjou region where the wine is produced without actually naming it, because this is not permitted for winegrowers who do not qualify for the AOC,

27 The odors and especially the flavors differ from those of conventional wines, because the wort has not been adjusted or manipulated by enological products.

28 Musk is a valuable ingredient in perfumery.

29 See the website https://dico-du-vin.com/.

30 For further details on his work, see Max Léglise, *Les méthodes biologiques appliquées à la vinification et à l'œnologie* (Paris: Le courrier du livre, 1994–1995).

31 Literally translated as "red cheeks."

in the Vin de France category. The wordplay here is based on the consonance in French between "les joues rouges" (red cheeks) and "Anjou rouge" (red wine from Anjou). Similarly, the contraction of *chenin haploïde*, "Chnaploïd," a non-reproducible and non-privatizable haploid chenin from Didier Chaffardon (near Angers), aims to denounce the farming of triploid oysters, which are genetically modified.[32] This neologism is a message to the political decision-makers who authorize GMOs. These messages also hint at the experiences and career journeys of these winegrowers, sometimes referring to very precise episodes. This is the case of one new winegrower, formerly a photographer and visual artist, whose first wine in 2020 featured an edited photo of her as a Native American in a headdress on the label and had the name "Sauvage" (savage or wild) (Figure 3.1). She made this choice for three personal reasons:

> My mother took this photo. We were coming back from a carnival and there was a Native American headdress in the truck. It happened very spontaneously. I edited it. You can't recognize me, but it reminds me of my mother, who died last year. Some people found the use of such iconography shocking, but I'm comfortable with it because it's not mockery. It's more like a double tribute. And there are two reasons for the name "Sauvage": When I did my first vinification tests, I pressed my grapes through crates, in my underwear, because I didn't have a press, it was very experimental, and what's more, I vinify outside, I don't have a place, it's a bit random, basic. Also, the track to my land is called "Le chemin de la sauvage" [the path of the savage/wild woman]! There's a coherence to it all.[33]

The different reasons she gives once again show how the concept of "wild," variable in both its perception and its usage (particularly, in the case of the French word "sauvage"), sways between evoking difficulty, freedom, autonomy, and sometimes even an element of chance (as is the case here).

The huge range of messages conveyed via the labels, which is constantly expanding as young winegrowers enter the profession, keen to take part in the labels game, has driven part of the conventional sector to refresh its communication strategy in recent years. Creativity can no longer be allowed to look like the exclusive marker of "natural" wines. Twisting the codes no longer seems to be the preserve of "natural" winegrowers, even if they are still more radical in their way of communicating. Well-informed observers can detect the differences. The disorder or hubris (to invoke the Dionysian myth) initially provoked by the "natural" movement, through both its practices and its way of publicizing them, seems to be seducing a sector confined within an image that is perhaps slightly too domesticated and policed, lacking rough edges. It is likely that marketing professionals

32 Some "natural" wine labels can also be serious.

33 Interview with Cécile Raynaud, August 2021. This winegrower also grows herbs and saved seeds.

Figure 3.1 Label "sauvage"

Source: Photo credit: Christelle Pineau

have understood that a little "rewilding" of part of the conventional sector, at least in terms of its image, could be beneficial and appealing. Have "natural" wines helped to disrupt the "wild" and "domestic" categories? Time will tell, since long-term observation will allow us to disprove or confirm this trend.

Conclusion: blending categories

Whether it is metaphorical or pragmatic, conscious or otherwise, the presence of the "wild" in the "natural" wine network makes sense insofar as it is the framework for these wines. Wild grapes, the original vines, embody the "wild" dimension and remind us that it was in fact "at the origin" of the "first" wines. In a contemporary context disoriented by damaging climate change, "wild" thought seems to act like a landmark, and to be defining for winegrowers who have chosen this path. It becomes a means to start a personal and collective journey in the vineyards afresh, following the route of the genealogy of wine. A resemblance can be observed with the recurring trends for a return to nature in so-called Western societies. Those described by Serge Moscovici in the 1970s echo the present of "natural" winegrowers, particularly on the concept of regeneration:

> Restarting means going back to the roots that supply sap to the numb senses, bloodless thoughts, and routine social relationships, obliging every individual, every community to wonder if they have been uprooted, if they have become an unviable graft or hybrid, transplanted onto different roots.
>
> (Moscovici 1974, 46)

Like the movement in the 1970s to reconnect with the Earth, "natural" winegrowers need to find vines whose roots go as deep as possible, both for the flavor of the wine and for it to be anchored in the environment. This way of "reconnecting [one's existence] with the Earth," of "making it concrete again" as suggested by Augustin Berque in his works on mesology (Berque 2010, 2015, 2018), updates the analysis of these reconnections with the Earth, in that it takes into account the climate emergency and the apostasy of modern dualism. Nevertheless, most of the global wine industry is still lulled by the promises of modernity. It clings to a production model built around synthetic chemical products, technology and yields. This part of the industry looks critically on the reverse evolution of "natural" wine production. The guardians of the dominant model reject the question of the "wild," pushing it back into the confines of the history of vines and wine. They do not wish to see the "wild" and "domestic" categories undermined. But while some are afraid to ease open the door to the "wild," others are experimenting to bring it (back) to life, even partially. Thinking about the "wild" can therefore be understood as an incantation, a palliative for forgetting or an attempt to ward it off. But studying and experiencing the "wild" also tell us about the environmentally non-invasive dynamic of research and adaptation to climate change established by "natural" winegrowers. In their context, the "wild" becomes a concrete tool for thought in the face of the programmed obsolescence of the all-powerful domestic. But rather than discarding one category, they incorporate it into the other, to try and take the best from both. In this melting pot, or this "gap between cultivated and wild," one of those "conceptual spaces and real places into which powerful demarcations do not travel well" (Tsing 2020, 283), combinations are created, which they hope will be fruitful. The blending of grape varieties is common among these winegrowers, according to their desires, the possibilities and the constraints. The blending of categories is common too: every day, they combine knowledge from the physical world and the academic one. The introduction of the "wild" into the "domestic" and vice versa, in varying proportions, seems entirely "natural." This blend of a wild utopia with a continuation of the domestic hints at a possible avenue for concrete reflection in the context of climate change.

References

Amiel, Christiane, Giordana Charuty, and Claudine Fabre-Vassas. 1981. *Jours de vigne. Les femmes des pays viticoles racontent le travail*. Terre d'Aude. Villelongue-d'Aude: Atelier du Gué.

Arnold, Claire, François Gillet, and Jean-Michel Gobat. 1998. "Situation de la vigne sauvage, vitis vinifera ssp. Silvestris en Europe." *Vitis 37*(4): 159–170.

Berque, Augustin. 2010. *Milieu et identité humaine. Notes pour un dépassement de la modernité*. Paris: Éditions Donner Lieu.

Berque, Augustin. 2015 [2000]. *Écoumène. Introduction à l'étude des milieux humains*. Paris: Belin.

Berque, Augustin. 2018. *Glossaire de mésologie*. Bastia: Éditions Éoliennes.

Dalgalarrondo, Sébastien, and Tristan Fournier. 2020. *L'utopie sauvage*. Paris: Les Arènes.

Dalla Bernardina, Sergio. 2011. *Le Retour du prédateur, mises en scène du sauvage dans la société post-rurale*. Rennes: Presses Universitaires de Rennes.

Descola, Philippe. 2013 [2005]. *Beyond Nature and Culture*. Chicago: University of Chicago Press.

Descola, Philippe, ed. 2018. *Les Natures en question*. Paris: Odile Jacob.

Gorz, André. 1988. *Métamorphoses du travail. Quête du sens. Critique de la raison économique*. Paris: Éditions Galilée.

Johnson, Hugh. 2020. *The Story of Wine: From Noah to Now*. Académie du Vin Library.

Larrère, Catherine, and Raphaël Larrère. 2015. *Penser et agir avec la nature. Une enquête philosophique*. Sciences Humaines. Paris: La Découverte.

Léglise, Max. 2007 [1990]. *La Vigne et le Vin entre Ciel et Terre*. Paris: Les Éditions du Vin/LeRouge&leBlanc.

Lévi-Strauss, Claude. 1990 [1962]. *La pensée sauvage*. Paris: Plon.

Mauss, Marcel. 2013 [1936]. *Sociologie et anthropologie*. Paris: PUF.

Moscovici, Serge. 1974. *Hommes domestiques et Hommes sauvages*. 10/18. Paris: UGE.

Néauport, Jacques. 1997. *Jules Chauvet ou le talent du vin*. Paris: Jean-Paul Rocher.

Pineau, Christelle. 2017. *Anthropologie des vins 'nature,' la réhabilitation du sensible*. PhD dissertation, École des Hautes Études en Sciences Sociales (EHESS), Paris.

Pineau, Christelle. 2019. *La Corne de vache et le Microscope. Le vin 'nature' entre sciences, croyances et radicalités*. Sciences Humaines. Paris: La Découverte.

Pineau, Christelle. 2021. "La vitiviniculture 'nature' et ses risques. De la liberté recouvrée aux actes de resilience." *Études Rurales 207*(1): 190–208.

Roupnel, Gaston. 1974 [1932]. *Histoire de la campagne française*. Terre Humaine. Paris: Plon.

Tsing, Anna L. 2020. *Friction, Délires et faux-semblants de la globalité*. Paris: La Découverte.

4 The wild nature of *vins nature*[1]

An oeno-centered counterpoint[2]

Léo Mariani

To add or not to add inverted commas: do we perceive the world through representations?

My aim here is to provide a counterpoint to Christelle Pineau's beautiful, very well researched chapter on the categories associated with *vins nature*, and the fluctuations of these categories. My comment will focus on an epistemic point, a divergence in opinion about the conditions of knowledge production, and the way wine and its actors are viewed. I feel the more encouraged to provide this comment as the author took a stance opposite to mine in her long introduction. There she declared herself in favor of using inverted commas with the word *nature*, inverted commas that I had elsewhere suggested to do without (Mariani 2021). As it appears, the punctuation marks in question are a point of contention that could indeed make for an interesting starting point for the discussion.

I understand that by placing the word *nature* between quotation marks – like many other terms such as "domestic" and "wild" – Christelle Pineau wanted to point to the polysemy of the word. In doing so, she, as the author, signals to the reader that the term in question does not designate what it might appear to designate literally – that is, she positions herself in a symbolic register. Necessary as this might sometimes be, the insistence on placing many words between inverted commas feels problematic to me. Moreover, this practice seems to reveal a type of ambiguous positioning, and it is in relation with this positioning that my comment is organized. In order to warn the reader not to take the categories the author uses literally, the author must first assume that he or she risks doing so, in other words that the reader spontaneously perceives the world through categories. In this case, a reminder that the words do not actually designate what they appear to designate might be indeed welcome. Another implication of this warning, however, is that

1 Translator's note: Although currently in use, the English term "natural wine" is subject to debate within the oenological community. Some argue that "natural wine" can be misleading, favoring "plain wine," "unadjusted wine" or even "no-additive wine" as more accurate translations of the French notion of *vin nature*. See www.winescholarguild.org/blog/the-great-debate-natural-wine-with-andrew-jefford-and-simon-j-woolf. Therefore, and also in light of the linguistic discussion of the French term *nature* in this text, I decided to keep the French term as such.

2 Translated from the French by Ioana Miruna Voiculescu.

DOI: 10.4324/9781003243496-9

the actors of the ethnography (the winemakers here) themselves are in fact oblivious of the artifice; unaware of the warning contained by the inverted commas, they remain automatically prisoners of appearances, trapped in a literal categorical interpretation of their reality. More generally, the overuse of inverted commas gives the impression that the world today, the observable world, resists description; hence, the need to constantly relativize it. As if there were an outside reality that existed by definition (i.e., nature) of which we could only make disconnected and fleeting representations.

When used as categories in a (modern) scientific inquiry, *natural*, *domestic* and *wild* do not need to be placed between inverted commas, because they have been specifically created – "purified" (Latour 1993) – to designate specific things and to order our understanding of the world. The actors have their own categories too. While these might sometimes seem to overlap with the scientific ones, since the words are the same, it is indeed rare that they designate the same things. And that is because the actors do not aim to be scientific, namely to use these terms with accuracy. Therefore, we, as researchers, cannot take the actors' representations to reference our own, or for that matter take the use of the word *wild* by the actors to be commensurable with the use of the same word by researchers. Moreover, we cannot pretend that these representations play the same role in organizing the experience of people as they do in organizing the experience of researchers. Describing and classifying are certainly ways to understand and know the world, but they are not the only ones. In failing to acknowledge that, we avow ourselves logocentric, conflating knowledge in general with a particular (scientific) form of knowledge.

Vins nature: turning the modern perspective upside down

Let us therefore start by overturning the assumption that people perceive the world through representations and that these representations are in themselves liable to describe the states of the world, on the model of natural sciences. In other words, let us not take the fluctuating meanings of words too seriously and focus instead on the phenomena they are meant to reference.

The French term *nature* can be used as both a noun and an adjective. As a noun, it describes an essence: it is the nature studied by biologists, an object which is real for them and, as a result, which they never place between inverted commas. This nature can be paired with the adjective *naturel* (natural), which conjures a degree of naturalness. Anthropologists of wine often place the adjective between inverted commas precisely to signal that the quality to which it refers is not fully accomplished – or, in other words, that it does not meet the scientific ideal. Consequently, the way people use this term is interpreted vis-à-vis an abstract definition and its corresponding (modern) model of knowledge. Yet I don't know of any winemaker who, by claiming that his or her wine is natural, means to say that the wine exists as such in nature, that is, it is produced outside any relationship with it. In practice, the adjective *natural* does not refer to the naturalness of wines but to a more subtle quality that deserves to be discussed, namely the absence of artificialization,[3] that is, the refusal to incorporate inputs,[4] as Christelle Pineau rightly observed.

This specificity is better rendered by the French (invariable) adjective *nature*, which is less equivocal than *naturel* (natural), despite the two being used interchangeably in practice. As an adjective, *nature* designates a pragmatic quality, an appearance without makeup or artifice. For instance, in French, yogurt is said to be *nature* when it has no flavoring added, or a person is said to be *nature* to emphasize her spontaneity, the fact that she does not care about conventions. Spontaneity is not assessed in the absolute or against an ideal fixed once and for all (Nature); it is instead assessed in relation to an experience, a phenomenological point of view. In this sense, one does not expect a *vin nature* to be natural, instead – just like one expects a person to "be without makeup"[5] – one expects it to demonstrate "sincerity."[6] The anthropologist should be prompted, therefore, to think outside the modern frame of evaluation, for, if the naturalness of a wine is not a relevant standard, then other ways to speak about it and those who produce it must be out there. A good starting point would be to ask a different set of questions.

To sum up, a *vin nature* is not a cultural *object*; it is not a matter *on which* representations and conventions, commercial, ideological or otherwise are projected but an honest, autonomous partner with *whom one* builds something. Surely, wine participates in cultural histories – it is produced, sold and consumed – but as an autonomous actor. It retains its freedom to assert its qualities in relations. If rewilding is defined – scientifically – as living outside society and conventions, then *vins nature* are by definition wild. They enjoy the ultimate freedom of living independently and of carrying this disposition to live freely, even imposing it on humans. While this does not prevent them from having relationships with humans, these relationships cannot be grasped through the wild/domestic frame.

Sulfur-free wines rewild the world around them

Let us now make an effort toward symmetrization, leaving behind for a while the anthropocentric point of view of representations and its underlying cognitive assumption, and consider how a wild natural wine acts on the world around it. To this end, let us use sulfur dioxide as a point of entry. Sulfur dioxide, or SO_2, is generally used as a wine preservative, stabilizing the musts and the drink by making them less dependent on oxygen. It also prevents the development of certain

3 On the other hand, a lower level of artificialization means a higher level of naturalness. But here the essence is longer placed in object position; it is indeed the consequence of a different purpose, to which it becomes subordinated.

4 Water included – which can be used to lower the level of alcohol if it is too high – despite it is being a natural product.

5 This description shows up a lot.

6 This, like the one in the previous note, is a term that I heard from the winemakers themselves.

components that human sensibilities find unappealing. Indeed, the mastering of this chemical component in the early twentieth century revolutionized the history of winemaking.[7] In France, it was at the root of a "conservative" paradigm, intrinsically linked with the opportunities offered by sulfur dioxide to stabilize wines and, in doing so, make them comply with human temporalities and designs. Guille-Escuret (1989) shows how the tradition (and the aesthetic) of *vins de garde* largely overlapped with the big Bordeaux bourgeois dynasties that found in this definition of the drink an opportunity to pass on an inheritance that only increased its value over time, a stock that could serve them to capitalize on their socioeconomic ambitions. A wine that can be kept is a wine available during the windows of opportunity, whether social or commercial. It is also a wine whose quality is easier to predict, "reliable" bottles that one can taste, for example, in south-western France before safely describing and selling them to American buyers.[8]

Choosing to stop using sulfur dioxide, the winemakers producing *vins nature* deprive themselves of an important ally. Giving up a fundamental means of control, they make their wines much less available for their own designs, much less predictable, more uncertain and more difficult to manage. By that, they stop defining wines as objects, as malleable supports on which they can inscribe their intentions and conventions before submitting them to the public's evaluation. And they also deprive themselves of the possibility to provide a guarantee for their wines, the way it is done for other products, in advance. They limit conservation and become constrained to find other faster outlets for their wines. In the absence of sulfur, an entire ecology of relations must be revisited and thought anew; the wines call for new forms of marketing, oenological techniques, storage methods,[9] assessment criteria (Mariani 2021) and forms of sociability that are shorter-lived than those of their cousins preserved with sulfur – more recreational, and more urban as well.[10]

Choosing to acknowledge and preserve the autonomy of their wines, the winemakers put them in a position to impose some of their qualities – qualities that

7 The use of sulfur in winemaking and oenology dates back to a long time ago. I am referring here to the mastering of the use of sulfur, which was achieved relatively late.

8 The links between the storage capacities and the development of social inequality were amply documented. They played a key role in the Neolithization of human societies, but they predate this process by long (Scott 2017) and persist to this day. It can be said that the nineteenth- and twentieth-century Bordeaux area provided a remarkably favorable ground for these processes in the form of technological innovations and social interests.

9 For example, *vins nature* don't fare well under Protected Designation of Origin standards, which are designed to stabilize specific oenological qualities and are therefore incapable to account for variability and diversity.

10 It is difficult to associate sharing a Bordeaux grand cru – too expensive and too difficult to drink during its young years because of the high tannin level and oakiness, both artifices that improve preservation – with the playful consumption that *vins nature* afford. Just as it is difficult to associate a wine tailored to patiently last throughout the years with the aesthetic of the ephemeral and the impermanent that *vins nature* call for.

they have gathered outside of the relationship – outside of the human. Such dispositions arising outside of any direct anthropic control[11] are wild by definition. They are brought out *of the wine itself* by native yeasts and bacteria. To the extent to which they constrain the humans to give up some of their expectations and social and cultural conventions, one might say that these dispositions specific to *vins nature* have the power to rewild in turn the world around them.[12]

Finally, adopting for a while, "the point of view of the wine" allows us to relativize the point of view of humans, and more specifically their role in rewilding wines. *Vins nature* is wild by definition. Because they tend to impart this quality to their surrounding world, they are as much a cause of the change as they are a consequence of it. They cannot therefore be said to be the product of a fashionable ideological quest, because they stem from a deeper-reaching, ontological reversal: the will of certain humans to give a different definition of what wines should be and of the place they should have within relationships. The object here is not the wine but the relationship.[13] In principle, this relationship is egalitarian, since it is established between equally autonomous entities. In practice, it does not rule out the development of forms of domination but it makes them more difficult to become established and perpetuated. In dealing with a *vin nature*, humans willingly force themselves to make a detour; that is, they avoid a prioritization of their designs and redirect anthropic pressure from the wine toward its ecology Less focused, this pressure is more difficult to apply to wines. This makes the relationship fundamentally more balanced, cosmopolitical by definition.

It is this relationship, and its quality and its maintaining, that the winemakers I worked with seemed to seek. They may not have used the scientifically right words to describe it – is that even possible? – but everything in their attitude and the daily diligence, they exerted to maintain their commitment communicated it.

On trust

Giving a place to other actors than humans in anthropological accounts is one of the major challenges for the discipline. One way of doing that is to focus on people who give themselves a more prominent place to the beings that they work with, not only in their accounts but, most of all, in their practices. Some winemakers do that by constituting their relationship with their wines around the radical choice to no longer artificialize them. In this respect, they do more than represent their

11 The wines are indeed controlled but indirectly, as their integrity and autonomy are preserved.

12 This explains why *vins nature* are generally described using terms that express their individualistic nature: "rebel," "original," "impolite," "unconventional," "punk" or simply "free."

13 I argue elsewhere that, at the origin of this choice, there must also be a will to reclaim a power to define and explore possibilities that agricultural modernity has largely preempted.

drinks differently – if indeed representation is even an aim for them. What they do is acknowledge the integrity of their wines, thus allowing them to play a more important, even causal role in human–wine relations. And, in doing so, they prepare the ground for a processual, open ontology, where uncertainty is a core value and rewilding a tropism.

In my ethnographic practice, I have made the choice to share my interpretations with the people and invite them to discuss my work, since I take the liberty to discuss theirs. In this context, I was often faced with the winemakers' criticisms, reproaching me that I tried to account for their practices with the help of representations that they found "caricature-like." At first, I thought that these representations were not the right ones and that I had to find new ones. Finally, however, I realized that they had themselves a hard time putting the essential, what mattered most for them into words. Often these difficulties ended with them inviting me to "see for myself," suggesting that I needed to participate, to engage in order to understand. In short, they thus confirmed that what deserved attention was not located in their representations – that is, in their minds – but in their wines and the relations that they maintained with them. By questioning my frame of reference, they asked me to embrace a different epistemology, which does not make the relations with the world conditional on words (which is not to say that this should prevent us from using words to speak about these relations).

Discussing the modern opposition wild/domestic in terms similar to my line of argument, Ingold (2000) recalls the relations between hunter and prey. Like in the case of wine, the prey is never entirely an object; it is never fully available and accessible to human designs. Now, hunter-gatherers do not try to cancel out this risk. They do not try to control nature but rather the relationship they have with it – in arguing this, Ingold cites Ridington. Hunter-gathers manage uncertainty indirectly; they choose to build trust rather than transform the world:

> That is why hunters attach such enormous importance to knowledge and its acquisition. This is not knowledge in the natural scientific sense, of things and how they work. It is rather as we would speak of it in relation to persons: to "know" someone is to be in a position to approach him directly with a fair expectation of the likely response, to be familiar with that person's past history and sensible to his tastes, moods and idiosyncrasies. . . . [T]he hunter does not transform the world, rather the world opens itself up to him.
>
> (Ingold 2000, 72)

Acknowledging the wild nature of their wines, winemakers make a similar choice, with similar implications. They establish a relationship based on mutuality and trust ("a fair expectation" as Ingold put it), a moral relationship between autonomous individuals that requires putting in place tailored forms of knowledge and judgments.

References

Guille-Escuret, Georges. 1989. *La Souche, la cuve et la bouteille. Les rencontres de l'histoire et de la nature dans un aliment: le vin*. Paris: Éditions de la MSH.

Ingold, Tim. 2000. *The Perception of the Environment. Essays on Livelihood, Dwelling and Skill*. London and New York: Routledge.

Latour, Bruno. 1993. *We Have Never Been Modern*. Cambridge, MA: Harvard University Press.

Mariani, Léo. 2021. "Cosmopolitiques vitivinicoles, ou pourquoi enlever les guillemets aux vins natures." *Essais. Revue interdisciplinaire d'Humanités*, Hors-série 6: 149–159. https://doi.org/10.4000/essais.7803.

Scott, James C. 2017. *Against the Grain. A Deep History of the Earliest States*. New Haven: Yale University Press.

Conversation 3. The revival of fasting

5 The promises of fasting

Between animal and primitive models[1]

Sébastien Dalgalarrondo and Tristan Fournier

In her article titled "Apostles of Abstinence: Fasting and Masculinity During the Progressive Era," R.M. Griffith (2000) shows how, in the early twentieth century in the United States, the promise of fasting was "retailored" by certain promoters to support a masculinist discourse: "a shriller gospel of masculine self-mastery and force." Fasting was no longer the preserve of women concerned about purity and repentance. Quite the opposite: it was becoming a technique of the body available to the conquering entrepreneur and the self-made man: "It was a crusade, moreover, in which food refusal played a key role, not simply among troubled (or fame-seeking) women but among highly enterprising middle-class men" (Griffith 2000, 632–633). The author thus shows that there is a strong political dimension to this discourse on the body. The promoters of fasting whom she analyzes sought to forge the modern entrepreneur who, armed with perfect health and flawless religious morality, would be able to handle the brutality of the market: "an eschatology of the body that was also a sacralization of the market economy. In and through the capacious practice of fasting, religion, masculinity, and capitalist dreams coalesced" (Griffith 2000, 603). Fasting was therefore part of the capitalist ethic and the productivist logic: "extreme food reduction appropriately suited the period's capitalist ethic of increased, efficient production" (Griffith 2000, 605). Our analysis fits into the perspective established by R.M. Griffith. We wish to show that the contemporary practice of fasting is based on a promise of "rewilding the self" that takes the form of a dual process. 1) A regressive process, a rewilding of the body based on a primitivist philosophy that aims to restore modern humans' physiological potential via an ancestral technique of the body: fasting. 2) A political process, a politicized rewilding of everyday life in the form of a radical experience of hunger aiming to challenge commercial routines and dietary habits, and to renew sensations and perceptions.

Fasting: a protean practice

Fasting, as a voluntary restriction of food intake, is central in many religions. It is therefore analyzed by specialists as a practice of self-control and self-transformation,

1 Translated by Cadenza Academic Translations.

DOI: 10.4324/9781003243496-11

and the expression of a frugal way of life (Hervieu-Léger 2017). Although religion is the ultimate domain of this technique of the body, fasting has taken on very different forms in recent history. An obvious example is political fasting, with the figure of Gandhi and hunger strikes. The fasting person is also a recurring figure in the world of music hall, seeking to thrill the audience through an exploration of the body's extreme capacities (Gooldin 2003). But beyond these forms, fasting can also be an everyday practice, as for certain Native Americans (Blumensohn 1933) who, when asked about the meaning of their regular fasts, did not know how to answer: for them, fasting was as inexplicable as hunger, defecation, or digestion. Finally, fasting as a health practice was promoted throughout the twentieth century by many schools of thought, from natural hygiene to naturopathy. In Europe, these different promoters of fasting managed to institutionalize the practice, for example, in Sweden, Germany and Russia. Clinics, which still operate today and are enjoying greatly renewed interest, offer therapeutic or preventive breaks. In Germany, for example, therapeutic fasting, practiced at the Charité – Universitätsmedizin hospital in Berlin, has been part of the normal array of treatments used by doctors for decades. Its use to treat certain illnesses is therefore publicly funded. The promise of fasting has recently become scientifically topical again thanks to the high-profile works of Professor Valter Longo, an expert on the biology of aging. Thus, as we will describe, fasting has become biomolecularized and is connected to the Promethean dynamic of transhumanism. In France, the recent development of numerous scientific controversies, particularly in the sensitive domain of cancer, continuously keeps the promise in the spotlight. The French biomedical institutions, concerned with containing the development of marginal therapeutic practices and channeling patients' hopes, emphasize that fasting has no proven benefits for human health (Rapport NACRe 2017). Meanwhile, the promoters of fasting are exploring the interstitial opening offered by the wellness, anti-aging and performance market.

Studying the French "wellness" fasting market

To analyze the contemporary practice of fasting, we will examine the recent development of the French "wellness" fasting market, in the standardized form of week-long group retreats. The vast majority of these voluntary deprivation retreats combine fasting with daily hikes. It is in this specific form that fasting practice has been spreading in France for around 20 years, in close connection with the growth of the alternative medicine and naturopathy market. It also exists in a specific dietary context, characterized by a process of *nutritionalization* (Poulain 2009; Scrinis 2013), a growing distrust of industrial food (Lepiller and Yount-André 2020), and an intensification of what is known as dietary reflexivity, meaning the growing recognition and consideration of the impact diet can have on health, the body and the environment (Fournier 2014). In order to understand what this redeployment reveals about the public's expectations and the ideologies at work, we will analyze this promise of wellness fasting in France by studying the content of the websites promoting the retreats. Our approach therefore belongs to the field of the sociology of expectations (Borup et al. 2006), which has shown the role that declarations about the future and strategic anticipation play in the launch of new scientific programs, the introduction of new technologies onto the market and the institutionalization of new practices. By choosing the term *promise* (Dalgalarrondo and Hauray 2015) to refer to this strategic actualization of the future, we will highlight the relational and circular dimension of the phenomenon. In fact, the performative efficacy of the promise is guaranteed only by its capacity to draw on shared aspirations and compatible imaginaries. This approach to therapeutic innovation aims to shed light on the legitimization processes involved. Beyond organizational phenomena, the focus is mainly on the cultural dimension, on the changing or shifting norms, on the symbolic rearrangements brought by these promises and on the ideological investments at work.

The first part of this chapter presents the fasting market in France, first from a socio-historical perspective by retracing the development of the "fasting and hiking" concept, then by characterizing the online offer. In the second part, we will pursue the idea of primitivism explored by R. M. Griffith, by analyzing contemporary fasting practice as a rewilding of the self.

1. The development of the "fasting and hiking" concept in France

1.1. The importation of a concept

The contemporary practice of wellness fasting in France, as promoted by many websites (which we will analyze later), was introduced in 1990 by Gisbert and Gertrud Bölling, a German couple living and offering horse riding in Drôme. The two *soixante-huitards*[2] (in their own words) attended a fasting and hiking retreat in Delémont in the Swiss canton of Jura in March 1990. Excited, they returned to France, gathered some friends from the organic farming community and organized a first fasting and hiking retreat the very next week, over Easter. This week-long retreat was inspired by the concept created in Germany by Christophe Michl, a teacher involved in the "marches for hunger,"[3] with whom the Böllings trained in the Black Forest. The protocol differed in two ways from that advocated by the natural hygienist school, which was very fashionable at the time, particularly due to the notoriety of Herbert Shelton, a naturopath recognized as its founding father. First, unlike the natural hygienists, who recommended strict water fasting, Michl and the Böllings were promoting Buchinger fasting, named after the German doctor considered to be the pioneer of therapeutic fasting[4] and who showed

2 Translator's note: The term *soixante-huitard* refers to a participant in or supporter of the demonstrations that took place in France in May 1968, or more generally to a social activist.

3 Inspired by the 1954 "Swedish march" – 11 Swedish doctors walking from Gothenburg to Stockholm (520 km) while water fasting to raise public awareness about fasting as a therapeutic practice – Christophe Michl took part in protest marches while fasting (the "hunger marches"), particularly for Third World autonomy. In 1984, he organized the first fasting and hiking retreat in Germany.

4 Suffering with rheumatoid arthritis, he tried fasting, with good results. He opened a first therapy center in 1920, and then, in 1953, he opened the famous Überlingen clinic by Lake Constance, with his daughter and his son in law.

that a small daily calorie intake (200 to 300 calories, corresponding to a fresh fruit juice and a clear vegetable broth) could alleviate the unpleasant effects of the first days of fasting (nausea, migraines, etc.). Second, whereas the natural hygienists encouraged complete physiological rest, fasting and hiking retreats combined fasting with daily physical activity (walking), claiming that this supports or even accelerates the body's "self-restoration" process.

Despite great popularity in Germany, fasting and hiking retreats struggled to take off in France. The Böllings promoted them via small ads in the local press and their stand at the Marjolaine organic exhibition in Paris. Demand was initially low but grew gradually, particularly thanks to media events. A first article published in the practical ecology magazine *Les 4 saisons du jardin bio* and written by a journalist who fasted with them in late 1990 brought the first retreat participants. Ten years later, following the publication of Gisbert Bölling's first work, *Le jeûne* ("Fasting") in 2000 and a promotional article in *Psychologie Magazine*, real demand emerged. Some of their retreats had up to 45 participants. After another decade, in 2011, the broadcasting of Sylvie Gilman and Thierry de Lestrade's documentary *Le jeûne, une nouvelle thérapie?* ("Fasting: A New Therapy?") on the French television channel *Arte* marked great enthusiasm for the practice. Today, the number of people who fast in France is estimated to 8,000 to 10,000 per year (according to the promoters whom we interviewed). Although demand seems to have significantly increased, the offer has also diversified considerably during this period: more than a 100 centers organizing fasting and hiking retreats can be inventoried online, across three main networks.

The first, in terms of both age and the number of members, is the Fédération Française de Jeûne et Randonnée (FFJR) (French Fasting and Hiking Federation), founded by its president Gisbert Bölling to bring together retreat organizers and promote this practice. Very quickly, dissenting voices emerged, including Eric Ledun, a former consulting firm manager who discovered fasting on a retreat and instantly saw an opportunity. He wanted to democratize this practice by dismantling its "hippie" image and emphasizing wellness rather than the concept of "therapy" or "treatment." The aim was both to avoid criticism for sectarianism[5] and to defend against accusations of illegally practicing medicine. By diversifying the offer (via personalization in terms of duration, practices and support) and by modernizing it (through the addition of wellness treatments and comfortable facilities), this network wanted to make fasting part of a more traditional personal development program offer. In 2008, he therefore created the Jeûne et bien-être (Fasting and Wellness) network, which he franchised. The eponymous magazine that complements this network can be analyzed to demonstrate the reframing. A third and more recent network emerged in 2010, in the form of a non-profit organization called VitaDetox. There are around 15 centers organizing fasting in this network.

5 Until 2009, fasting and hiking retreats were monitored by Miviludes (a French government agency that monitors sectarian movements) for risk of "sectarianism."

1.2. *The fasting web*

Thanks to our research, we identified over a 100 centers with an online presence, the vast majority of which are approved by one of the three networks described earlier. After, first, smoothing our sample, we decided to focus on the first 60 websites in order to map the offer. Table 5.1 presents their main characteristics.

Table 5.1 Analysis of websites (n = 60)

Retreat offer online		
Centers (n = 60)	Profile	- Fasting and hiking sites (n = 30) - Naturopathy sites (n = 12) - Wellness centers (n = 9) - *Gîtes* (vacation rentals) and unusual accommodation (n = 5) - Yoga associations (n = 4)
	Affiliation	- FFJR (n = 32) - Jeûne et bien-être (n = 11) - VitaDetox (n = 10) - None (n = 7)
	Price	€680 on average (from €350 to €1,300) for a one-week, all-inclusive retreat (support and accommodation based on one person in a double room, treatments not included)
	Presence of medical personnel	- General practitioner (n = 6) - Nurse (n = 3) - Nutritionist (n = 1) - Pharmacist (n = 1) - None (n = 49)
	Location	Retreats all over rural France, but a particularly high number in Brittany, Provence, and the French Basque Country.
Organization	Phases	- One week of "food descent" (progressively and chronologically removing meat and fish, followed by tea, coffee and chocolate, then carbohydrates, so that the participant is eating only fruit and vegetables on the day before the retreat) - Six-/seven-day retreat - One week of "food reintroduction" (progressively and chronologically adding food groups back into the diet in reverse order)
	Typical day	- Drinking organic fruit juice together, discussion - Waking up the body (yoga, daoyin, etc.) and/or meditation/sophrology - Hiking (for two to four hours) - Free time, with access to: o Facilities (swimming pool, jacuzzi, "bol d'air Jacquier" oxygenation treatment, etc.) o Treatments (massages, reflexology, etc.) (not included in the price) - Eating an organic clear broth together, discussion - Seminars and/or workshops (nutrition and naturopathy)

Although prices vary widely (from €350 to €1,300 per person for a one-week retreat, including accommodation) and the facilities and comfort levels of the centers vary considerably, we can observe a certain homogeneity in terms of the schedule and organization. Each retreat lasts a week, is preceded by a week of "food descent" (during which the participant progressively removes food groups from their diet at home, until they are eating only fruit and vegetables the day before the retreat) and followed by a week of "food reintroduction" (progressively adding food groups back into their diet in reverse order), with the first meal prepared and eaten together on the last day of the retreat. A typical day is organized as follows: 1) drinking organic fruit juice together for breakfast, an opportunity for participants to talk and reflect about their sensations; 2) waking up the body (yoga, qigong and daoyin) and/or meditation/sophrology; 3) hiking (for two to four hours); 4) free time in the afternoon, with access to the facilities (swimming pool, jacuzzi, "bol d'air Jacquier" oxygenation treatment, etc.) and to treatments (massages, reflexology); 5) eating an organic clear broth together for dinner, an opportunity for further discussion; and finally, 6) a seminar and/or workshop (nutrition and naturopathy).

1.3. A *triple rhetoric*

By analyzing the websites, we can identify three major rhetorical trends in the promotion of fasting and hiking retreats in France: a critique of (dietary) "modernity," an emphasis on the value of (human) nature, and an invitation to take a (transformative) break.

THE CRITIQUE OF (DIETARY) MODERNITY

Fasting is presented as a chance for a person to (re)gain control over their diet, their body and their everyday life, at a time when "modernity" is accused of all evils. "[Fasting is] an intense bodily repair and cleansing session, . . . a process that can be seen as a way to rebalance in the face of our dietary excesses, our Western lifestyle, and all the resulting problems" (Jeûne in Landes website, accessed 19 February 2022).[6] The premise is that we eat too much, badly, and out of habit. A fasting and hiking retreat is therefore seen as a way of escaping this insipid "modernity" and giving new value to what we eat and new meaning to our lives. However, a retreat in which participants do not eat for a week and attend daily nutrition seminars seems a long way from the "French dietary model" with its focus on the gustatory, hedonic and convivial dimensions (Mathé, Tavoularis, and Pilorin 2009; Poulain et al. 2010). This emphasis is particularly embodied by the recent inclusion of the "Gastronomic meal of the French" on the UNESCO

6 Translator's note: Unless otherwise stated, all translations of cited foreign-language material in this chapter are our own.

Intangible Cultural Heritage list. It therefore seems necessary to examine what lies behind this promise and whom it targets. We might hypothesize that these retreats are the reflection of a *nutritionalization* process (Poulain 2009; Scrinis 2013 and of increased dietary reflexivity, given that the demands of the healthy and sustainable consumer are highly influential for contemporary diets (Dubuisson-Quellier 2016). Permeability to these demands is nevertheless socially determined (Fournier 2014). The budget required (at least €350 per person for a week) creates an initial social sorting effect. Paying to go without food for a week is part of a distinctive phenomenon (Bourdieu 1996) that is particularly active in our time of overabundant food supplies. Beyond the solely economic aspect, these fasting and hiking retreats explicitly target people who are already aware of the effects of diet on health, well-being and the body, with a critical perspective on "dietary modernity": "After fasting for a few days, you feel light and full of energy, all your senses are sharpened, and your brain is working much better" (Jeûner pour sa santé website, accessed April 2018).

EMPHASIZING THE VALUE OF (HUMAN) NATURE

Through his experimentation underpinning the work *Walden, or Life in the Woods*, Henri David Thoreau (1854) wanted to reconnect with nature to find *his* nature. This double level is perfectly identifiable in the rhetoric of the websites analyzed. Some arguments follow the theme "Be revitalized by nature's abundance" (Les sentiers buissonniers website, accessed 19 February 2022), while others claim that "Fasting is our true nature" (Jeûner côté mer website, accessed 19 February 2022). The rhetorical goal is to reconcile nature and human nature by presenting the causal link between them. Immersion in a rural setting and hiking, combined with experimenting with hunger, purportedly help participants to develop greater everyday attention to the body and mind, and to "find themselves" or even (re) discover themselves. Moreover, the idea is to surrender to "nature," because it is always seen as beneficial (Lepiller 2013), whereas artifice is seen as disturbing, particularly when it comes to food (Fournier and Lepiller 2019). "I invite you to do nothing for once and just surrender" (Jeûne et détox en Auvergne website, accessed April 2018). Nature and human nature are entangled and amalgamated here, and fasting and hiking retreats are presented as ways to reconnect with this nature.

A BREAK AS A GATEWAY TO TRANSFORMATION

The third type of rhetoric used is that of a break. It is omnipresent on the websites, on two temporal levels. On one level, it is the break itself that is emphasized, as a lived experience, or even an indescribable one:

> Those who have fasted before know it, first-timers will discover it, and I almost envy them! The first fast is undoubtedly the most unbelievable, because the

> experience is so powerful. . . . Despite this fabulous promise, some people still hesitate.[7]

It is made possible by participants being away from home: this is a retreat. It is a shared experience and the organization is delegated, but it takes place away from the hustle and bustle, obligations and habits of everyday life. The "fasting and hiking" retreat is like a form of self-experimentation, supervised and advised by expert fasters. This particular space-time has its own rhythm, rituals (consuming broth together in the evening) and challenges; particularly, the physiological challenges of the "ketonic phase"[8] (Gueguen et al. 2014) on around the fifth day of fasting are said to help participants discover new sensations and emotions: "Our body's adaptability is an extraordinary thing, which we don't notice in normal times" (Jeûne et détox en Auvergne website, accessed April 2018). On the second level, the break allows a fresh start and works like a gateway back to willpower. It is seen as the trigger required to change "unhealthy" practices with the possibility of "cleansing" everything to start from scratch: "fasting is a simple way to completely cleanse the body, wipe the slate clean, and make a fresh and positive start" (Jeûne Vitalité website, accessed April 2018). The idea is to relearn how to eat and live differently. The fasting break is therefore a time of experimentation, allowing subsequent better regulation of one's appetite and improved control over one's way of life.

2. Fasting as a way of rewilding the self

2.1. *Animals as a model*

Animals, particularly mammals, regularly feature in the contemporary discourse of those promoting fasting. In the strictly French context, the works of Yvon Le Maho, ecophysiologist at the Centre national de la recherche scientifique (CNRS) (French National Centre for Scientific Research), are regularly used to substantiate this animal-based argument. His works helped to demonstrate that the remarkable fasting capacities of penguins were shared in the animal kingdom, including by human beings. Therefore, the scientific marketing used by promoters of fasting regularly states that any person in good health can fast for around 40 days. This emphasis on inter-species biology as a way of rediscovering our otherness, a source of biomimicry, is a metaphorical invitation to rewild ourselves. Fasting is therefore

7 First sentence of the editorial in the magazine *Jeûne et bien-être*, signed by the editor in chief: Christine Bory-Huyghens, "Alors, prêt à jeûner?" *Jeûne et bien-être*, no. 3, January 2018, 3. Available at: www.flipsnack.com/jeuneetbienetre/magazine-je-ne-bien-tre-2018.html.

8 The ketonic phase starts on around the fifth day of fasting and can last several weeks depending on the physiology of the individual. Human adaptation to prolonged fasting is characterized by our capacity to save the body's proteins. Energy substrates are therefore primarily provided by lipolysis (the breakdown of fats) and the ketone bodies produced have the particularity of being able to directly fuel the brain. Many people experience an altered state of consciousness during this stage.

an experience, a scientific self-experiment and a biotechnological intervention "with bare hands," allowing the individual to optimize their body by reactivating their "true nature": "I invite you to try the extraordinary experience of fasting, to rediscover this innate capacity that lies dormant in every one of us. Fasting is our true nature" (Jeûner côté mer website, accessed April 2018). The construction of fasting as a scientific object and a way of exploring one's physiological capacities is strengthened by the recurring mobilization of research fronts, particularly the work of Valter Longo, a professor of gerontology at the University of Southern California, made famous by his publications on calorie restriction. In the 2000s, he demonstrated that reducing the food intake of laboratory mice improved their health and increased their lifespan. Moreover, he is at the origin of hopes about the possibility of fighting cancer or the side effects of drugs by fasting. His most recent publications concern experiments on humans, which aim to prove the efficacy of a particular fasting cycle (five days a month for three months) in combating chronic conditions associated with aging. Valter Longo makes fasting a universally available tool for biological regeneration, meaning that it rides on the back of the flourishing anti-aging promise (Dalgalarrondo and Hauray 2015). Here, the animal model serves as an ideal and practical horizon, allowing promoters of fasting to trace the path that "impaired," "degenerate" humans must follow.

Here too, a little historical perspective demonstrates the continuity of the animal-based argument. In the early twentieth century, Macfadden, one of the emblematic promoters analyzed by R.M. Griffith, regularly evoked animals that fast to convince readers of the universality of this self-healing mechanism and to reassure them. According to this argument, civilized humans needed to rid themselves of all artificiality by using fasting to recover the strength of the wild animal:

> "Civilized man" needed to throw off the artificial tastes and cultivated habits accrued over time in favor of the practices observable in animals and so-called "primitive" people. . . . This shift would restore the lost strength and vitality that every man supposedly coveted, enabling him to conquer all that was debilitating and feminizing about modern civilization.
>
> (Griffith 2000, 608)

Doctor Shelton, another American figure in the history of fasting analyzed by Griffith, devotes an entire chapter to fasting among "lower animals," in which he draws on the paradigmatic case of illness to normalize fasting and present it as natural. Wild and domestic animals become models, the keepers of a technique forgotten by modern humans:

> I need but devote little space to a discussion of what every one [*sic*] already knows; namely, that the sick animal refuses all food. The farmer knows that his "foundered" horse will not eat – is "off his feed," as he expresses it. . . . In fact, all nature obeys this instinct. Thus does nature teach us that the way to feed in acute "disease" is not to do it.
>
> (Shelton 1934)

Shelton emphasizes the instinctive dimension of fasting and sketches an original history of human biology:

> Fasting above all other measures can lay claim to being a strictly natural method. There can be no doubt that it is the oldest of all measures of meeting those crises in the organism called "disease." It is much older than the human race itself since it is resorted to instinctively by sick and wounded animals.
>
> (Shelton 1934)

Macfadden also sees wildlife as a source of inspiration and definitive proof: "serious sickness prompts all animals to fast. Wounded deer will retire to some secluded glen and starve for weeks together" (Macfadden and Oswald 1900, 40). Thus, the practice of fasting incorporates an original becoming-animal[9] in the form of an alliance, a new zone of exchange and assemblage with the animal world and its biological potential.

2.2. The hunter-gatherer as a model

Fasting is presented as time out, a liberating journey primarily based on renewed attention to the body and the establishment of a "dialogue," as shown by this representative account:

> My experience of facilitating fasting retreats has brought me many surprises myself: when I accompany the hikes, I eat in the morning before leaving and in the evening when I get back. After all, I can't fast for thirty-five weeks a year. But I do thirty-five weeks of a sort of Ramadan. . . . In winter, we finish hiking before dark, at around 3 or 4 p.m. In summer, we sometimes have a long siesta to let the heat pass, and we might even get back after 5 p.m. I had to talk to my body like you might talk to a good friend, to explain the reasons for the delay. But now, I can do anything: whether I eat or not, my body doesn't even protest. It just trusts me. There are no obligations anymore. I'm free to do as I like.
>
> (Bölling and Bölling 2017, 15)

In some discourses, taken to a deeper level, this dynamic of the journey becomes a real exploration of the self and of one's capacities to adapt. Promoters of fasting regularly evoke the demands of the human being and of human biological evolution, along with the theme of survival as the way of life of our ancestors. Fasting is therefore described as an experience that allows us to reconnect with

9 The concept "becoming-animal" was proposed by Gilles Deleuze and Félix Guattari. See the reference list for an English translation of their work: *A Thousand Plateaus: Capitalism and Schizophrenia* (2005).

Paleolithic humans, who trod lightly on Earth, in harmony with nature. Here, we encounter the thinking seen in evolutionary medicine, evolutionary nutrition and even evolutionary psychology, scientific fields that theorize the question of "mismatch diseases" (MMDs), "disharmony" or "caveman syndrome." According to these theories, the modern individual is handicapped, degenerated and weakened by their inability to live as their innermost nature dictates. The promise of "paleo" diets is based on the same primitivist principle, trying to reconstruct the original environment of the hunter-gatherer and getting as close as possible to their way of subsisting.

This primitivist theme is particularly prominent in the documentary *Le jeûne, une nouvelle thérapie?* ("Fasting: A New Therapy?"), which contributed to the revival of the practice in France. The hypothesis of a growing disconnect between human biology and our environment and modern lifestyle is extensively discussed in the documentary. The situation of abundance is presented as a historical anomaly in the long history of human evolution. Effectively, not fasting is presented as a suboptimum use of the body, or even a pathological state, as suggested in the documentary by Andreas Michalsen, Head of Department at the Charité – Universitätsmedizin hospital in Berlin: "It is not surprising that the body experiences difficulties if it does not fast, when it eats constantly. Our genetic heritage seems less adapted to this situation than to fasting." Therefore, a fasting retreat is presented as an opportunity for personal *scientific exploration* of a human potential: "Fasting is not a trend, it is a powerful technique that we are rediscovering, but which our body has not forgotten" (Les belles humeurs website, accessed 19 February 2022).

It is interesting to note that this contemporary "mismatch" theory appears, in another form, in the discourse of the fasting promoters analyzed by Griffith. To justify the practice of fasting, they regularly argue that contemporary living causes a loss of vitality. They offer a way of regaining the perfect health of our "ancestors" (a vague concept used to refer to pre-civilization man, the classical figure of the noble and strong savage). Diet is presented as one of the main causes of the "degeneration" of modern humans, the theory being that the organism is burdened by the need to evacuate unhealthy substances, toxins and above all excesses. According to Macfadden, Americans are

> actually eating themselves into their graves. . . . They eat, not to nourish the body, but merely for the pleasure of gourmandizing. The result is that from two to five times as much food passes through the alimentary canal than is necessary to maintain weight and strength, and mind and body are actually weakened by the strenuous efforts made by the system in endeavoring to rid itself of this excessive amount of food.
>
> (Macfadden and Oswald 1900, 82)

Modern, urban life is criticized for its pathogenic effects on the body and mind: "Who can doubt that modern living tends to dull the mental powers? Especially our national drug addictions and our almost universal overeating tend to reduce mental abilities" (Shelton 1934). In Shelton's view, the senses and instinctive reflexes (which include fasting) are also "dulled" by modern civilization.

> There can be no doubt that all of man's special senses are more or less dulled or weakened by civilized life and by his "disease" and degeneracy. In fasting, without the recorded exception of a single case, the senses are remarkably improved.
>
> (Shelton 1934)

The goal is to regain the supposedly perfect health of our ancestors, to once again be able to fully express our "natural heritage":

> It [the book] will teach you that weakness is a crime – that it is the result of plain, easily avoided causes – that if your body is weak, or diseased, there is not the slightest excuse for remaining so – that health and strength of a high degree is the natural heritage of man and woman.
>
> (Macfadden and Oswald 1900, 7)

Thus, the idea of instinct is frequently evoked, and those who fast are invited to relearn to interpret "the language of your sanitary instincts" (Macfadden and Oswald 1900, 11).

3. The politics of self-rewilding

3.1. *Asceticism*

Finally, the contemporary practice of fasting can be seen as an original form of resistance to dietary modernity. Analyzing this practice of breaking free from the norm takes us into the infra-political register (Scott 2008). The fasting retreats aim to make participants independent so that they can repeat the experience in the privacy of their own kitchens. The promise of fasting therefore allows each initiative to reproduce a practical criticism of the dietary order, away from the gaze of the state and commerce. This political rewilding of everyday life is primarily based on adopting a temporary but strict asceticism. By practicing and experiencing fasting, participants want to test and improve their self-control. Therefore, fasting is like a limited-term experiment, an escape allowing the participant to begin a reflection on the effects of abundance in our consumer societies. Unsurprisingly, the contemporary promise features the secularized incarnation of this asceticism: the athlete. According to Peter Sloterdijk (2014) the athlete corresponds to this ideal of advancement through effort. Sloterdijk hypothesizes that we are seeing a powerful return of the practice of asceticism in our secularized societies. Thus, modernity resembles a new anthropotechnic regime that has "secularized and collectivized the practicing life by breaking the longstanding asceticisms out of their spiritual contexts and dissolving them in the fluid of modern societies of training, education, and work" (Sloterdijk 2014). The practice of fasting is part of this massification of asceticism. Sport, as a laboratory and experimental context for human optimization, is like the great media *gymnasion* of anthropotechnics: "the methods of mental and physical practising by which humans from the most diverse cultures have attempted to optimise their cosmic and immunological status in the face of vague risks of living and acute certainties of death" (Sloterdijk 2014). The evocation of the emblematic figure of Yannick Noah (a high-profile French tennis player in the 1990s) in the "fasting and hiking" guide published by the FFJR in 2017 falls within this secularization of asceticism, in aid of an opening onto the concept of excellence:

> Journalist's question: You're starting your seventh day of fasting. Is it a difficult ordeal? Answer: Normally, I fast in spring and early autumn. It really does me good. I cleanse myself, I rid myself of the stress from the Davis Cup and the fatigue from traveling. What's more, I had started smoking and drinking again. I needed to quit completely, and it's done! The first two days are hard. After that, it gets gradually better, you feel good, all your senses are alert. . . . Personally, when I saw it through to the end, it felt like a victory over myself. We know that fasting is linked to spiritual exploration. We see it clearly with Ramadan among Muslims or in other religions. Gandhi, whose hunger strikes are famous, subsequently fasted for one day a week. In short, fasting takes us closer to the core of things and therefore to ourselves, to the inner voice, to the cosmos, you might say.
>
> (Bölling and Bölling 2017, 22)

This high-level athlete presents fasting as an exercise of mastering one's desires that helps to physically "cleanse" the body, awaken the mind and spirit, regenerate the senses and cultivate psychological strength. The athlete, who is a secularized embodiment of the ascetic and their "feats," demonstrates the possibility of self-production through scrupulous practice. This mirroring between spiritual and physical exercise is perfectly described by philosopher Pierre Hadot:

> Underlying this conviction is the parallelism between physical and spiritual exercises: just as, by dint of repeated physical exercises, athletes give new form and strength to their bodies, so the philosopher develops his strength of soul, modifies his inner climate, transforms his vision of the world, and, finally, his entire being. The analogy seems all the more self-evident in that the *gymnasion*, the place where physical exercises were practiced, was the same place where philosophy lessons were given; in other words, it was also the place for training in *spiritual* gymnastics.
>
> (Hadot 1995, 102)

The presence of the expression: "A healthy mind in a healthy body" (Hygea website, accessed April 2018) on one of the websites analyzed offers a different way of explaining the centrality of the body in the promise and the exploration of new forms of organization between body and mind, with the power of the ancient message and Olympic history: "A highly parasympathetic matter. Fasting, a rest for the body . . . and the mind."[10]

3.2 *Pleasures of asceticism*

Asceticism is closely linked with the concept of pleasure. First, there is the pleasure of the exercise itself and of a practice that gives an increased feeling of control. However, outside of pathological distortion, represented here by the emblematic figure of the anorexic, asceticism as an exercise has a more elaborate connection with epicureanism than surface appearances might suggest. "Fasting is also a way of reconnecting with your senses, for increased pleasure" (Osez jeûner website, accessed 19 February 2022). The practice of fasting, whatever its aim, has often been seen as a way of rediscovering the simplicity of everyday pleasures, particularly those related to food (Griffith 2000). By this logic, fasting provides an opportunity to escape habits and the numbing of the senses that they can cause. This epicureanism is present in many artistic or literary evocations of the moment when a religious fast is broken. Beyond the pleasure associated with dietary control (Dupuy 2013), the person eating also experiences renewed and deeper pleasure thanks to the voluntary effort made. Fasting therefore provides a way of sharpening the senses again, beginning a new relationship with food in which the search for taste and diversity replaces the "vulgar" pleasures of abundance. We can observe a desire for true products and a gastronomic relationship with food: "Fasting with us is epicurean, gastronomic, and organic" (Bio Alcaline website, accessed 19 February 2022). "Fasting is also a way of reconnecting with your senses, for increased pleasure . . . it is also a convivial moment" (Osez jeûner website, accessed 19 February 2022). And fasting is presented as just one stage, a founding moment that helps participants to reconfigure their desires, allowing them an opportunity to rediscover taste, to feel truly hungry again. They therefore undergo the common experience of the thirsty walker who, after many long hours, rediscovers the simple pleasures of water and its thirst-quenching powers by a bubbling spring. R.M. Griffith has clearly demonstrated the many facets of this quest for pleasure, particularly sexual pleasure, among nineteenth-century fasting doctors: "indebtedness to Dewey's influence, many highlighting the fast as an instrument of efficiency that would bestow riches galore. And not only riches: indeed, the faster harnessed the power to obtain practically any desire" (Griffith 2000). This disciplined way of living, in the words of Benjamin Constant cited by these doctors, provides a "power of will."

10 Translated title of the magazine article "Une affaire très parasympathique. Le jeûne, un moment de détente pour le corps . . . et le mental," *Jeûne et bien-être*, no. 3, January 2018, 10.

3.3. *A silent revolution*

Finally, a fast can be seen as a silent protest, meaning that every individual has the opportunity for domestic resistance. The kitchen, with its cupboards and food supplies, becomes a space where one's life can be politicized, in the everyday context (Pruvost 2021):

> Those familiar with fasting use it to sharpen and focus their mind for important decisions. For example, Lanza del Vasto fasted for 40 days before launching the Larzac campaign. Gandhi did the same before attacking the English in a nonviolent manner. These fasts greatly contributed to media coverage of the campaigns: a public fast always hits the headlines, but fasting also brings lucidity to the faster, which helps them achieve the goal they have set themselves.
> (Bölling and Bölling 2017, 14)

Collective learning of the practice is just one step, as G. Bölling states using the metaphor of the "swimming instructor": fasting is a technique that is easily learned and is therefore destined to be passed on. It can be used at home, meaning that individuals can regularly demonstrate a desire for "de-growth," breaking free from the flow, and rising momentarily above contingencies. It is the relationship to overabundance that is politicized here:

> For several decades, we have been living in a permanent state of abundance, but our body maintains the ability to "store" excess calories. This is why the ordinary European is almost always overloaded, heavier than their ideal weight, and often even obese. The body must support this unnecessary weight.
> (Bölling and Bölling 2017, 4)

Fasting offers a revolution *in and of* everyday life, a way of deconstructing the feeling of hunger, our habits and our mental hygiene: "'Being hungry' takes on a whole new meaning: It indicates not the immediate need for food, but simply the HABIT of eating" (Bölling and Bölling 2017, 28). Fasting is promoted as a way of combating the harmful lifestyle that is imposed on us every day:

> Overabundant, unnatural food and insufficient nocturnal elimination due to decreased sleep time are factors that cause "clogging." The body, overwhelmed by toxins, becomes weakened. But clogging = slower cellular function = accelerated aging = biological age higher than the real age. The goal is therefore to live not only as long as possible, but also in good health![11]

11 "Le jeûne, un soin anti-âge," *Jeûne et bien-être*, July 6, 2021, accessed April 2018, https://jeune-bienetre-magazine.fr/le-jeune-un-soin-anti-age/.

Fasting is presented as the catalyst for a time of introspection and reflection, a political questioning of everyday life (Lefebvre 1997). It is seen as a *ruse* in the face of consumer society, an act of "cultural poaching" in the words of Certeau (1990), an anthropotechnic bricolage, a tool for resistance similar to I. Illich's advocacy of "techno-fasting": he notes that "ascesis is the old word for training, for repetition. I would say what is required is a word which is difficult to pronounce today – virtue" and that "the certainty that you can do without is one of the most efficacious ways of convincing yourself, no matter where you stand on the intellectual or emotional ladder, that you are free" (Cayley 2005, 102, 228). Fasting is presented as a momentary escape from the everyday flow, in order to rediscover the liberation of letting go:

> Cultivating the art of letting go. . . . Our retreats help participants to take the step back we need in our lives. For both body and mind, fasting is a momentary escape from the "system" and allows us to regain some autonomy and therefore free will.
>
> (Jeûne et bien-être en Vercors website, accessed April 2018)

Fasting is therefore seen as a political act, in which stepping back from the world and its pollutions helps to develop a person's free will. "We fast without food, but also without the pollutants that poison us (tobacco, drugs, medication), in an unpolluted environment . . . with no television" (Jeûne in Landes website, accessed 19 February 2022). At its most extreme, this quest for freedom can take the form of "breatharianism." In this case, the technique of fasting becomes a mystical practice, and its mastery by certain people supposedly allows extraordinary modes of existence:

> This subject would not appear in this little guide if there weren't an Australian woman, Jasmuheen, who has not eaten for years and who travels all over the world to speak about it. Her line of argument is very simple: just as vegetarians might say that we ourselves can eat the plants that feed animals, she claims that we can be sustained directly by the energy, water, and light (prana, chi) that allow plants to grow. Her teaching in Paris convinced me so fully that I started "the process" for eleven weeks, without losing weight. . . . In comparison, our one- to three-week retreats seem rather outdated! Incidentally, Marthe Robin lived in Drôme. . . . But don't worry: you come to us to fast.
>
> (Bölling and Bölling 2017, 25)

Being "sustained by light": this is the political objective of absolute de-growth taken to its extreme. The evocation of breatharianism reflects the "politico-mystical" anchoring of the practice of fasting, admittedly on the fringes of the discourse of the promise but in a hyperbolic form conducive to the construction of a hierarchical scale that ranks those who follow these practices according to the degree of "liberation" or "initiation" achieved. Living on energy, water and light is presented as the ultimate, mystical stage in the rewilding of the self.

Conclusion: fasting as transhumanism "with bare hands"

The analysis of the contemporary practice of "wellness fasting" provides an archetypal example of a desire to improve one's body via an intellectual and practical movement of "back to," a regressive process allowing a person to regain their "original" vitality. This voluntary abstinence corresponds to the possibility of rewilding the self that we describe as "transhumanism with bare hands" (Dalgalarrondo and Fournier 2019b): a biotechnological honing, a process of "self-optimization" (Dalgalarrondo and Fournier 2019a), through the exploration of a "primitive" technique of the body. The hope of the biology of aging is combined here with a desire to reconnect with our "true nature," with the optimal physiology of the Paleolithic faster-gatherer, and with our spiritual lives. Fasting is seen as an exploration of the self and of our capacities for physiological and political resistance, a momentary liberation from the feeling of hunger with the aim of rediscovering the ways of a frugal, simple, minimalist life. These fasting retreats are like a new consumption of *spiritual exercises*: "The quest for self-realization, final goal of spiritual exercises, is well symbolized by the Plotinian image of sculpting one's own statue" (Hadot 1995, 102). Fasters shed the excess to discover and perfect themselves (Zask 2010). Professor Andreas Michalsen, in charge of therapeutic fasting at Berlin's Charité – Universitätsmedizin hospital, expresses it as follows: "we see it as surgery without a scalpel" (Lestrade 2015, 86). It is about more than just following a diet or adopting an aesthetic-moral position. People fast in the hopes that the exercise itself will allow them to experience self-transformation. Fasting is seen as a way to do more with less.

To a large extent, the performativity of this promise comes from its assembled form (Ong and Collier 2008) with its evolving boundaries. This assemblage allows the promise to extend into the public and commercial space by adapting to different audiences. It is an assemblage that operates like a tiled roof, each tile/argument overlapping with those around it: the scientific argument legitimizes the promise and gives it credibility, but this biological science, through its anthropomorphic effect, also helps to fuel and solidify the political dimension of fasting. The body's return to nature is connected to a political commitment, a challenging of abundance, and fasting becomes a necessary asceticism for better adaptation to our "pathological" societies. This assemblage therefore allows a sliding between different positions. Thus, the alternation between alternative discourse and scientistic discourse, identified in all the material collected, makes it possible on the one hand to legitimize alternative practices (the status of fasting and hiking retreats in itself), because they are supported by modern science (even if the publications in question are peripheral or still controversial). On the other hand, it also provides an authorization to discuss physical and/or mental performance (linked to the ideology of optimization), because the words come from an alternative position, which rids them of this ideological load. Finally, in this original assemblage, we can identify a hybrid configuration that echoes the recent works of philosopher Francis Wolff (2017), in which he presents human identity as being caught in a tension between two extremes: the immortality of the gods and the irrationality of animals. Yet he claims that this tension is lessening, with transhumanism bringing us closer to immortality and animal studies showing us the blurring of the boundaries between humans and animals. The contemporary practice of fasting is also caught up in this tension and allows us to understand the assembled imaginary (Jasanoff and Kim 2015) of the *rewilded being*, at the crossroads between transhumanism (a process of self-enhancement through optimization) and "naturalism" (reconnecting with nature and with *our* nature).

References

Blumensohn, Jules. 1933. "The Fast among North American Indians." *American Anthropologist 35*(3): 451–469.

Bölling, Gertrud, and Gisbert Bölling. 2017. *Jeûne et Randonnée. Petit Guide Pratique*. Accessed June 6, 2008. www.jeune-et-randonnee.com.

Borup, Mads, Nik Brown, Kornelia Konrad, and Harro Van Lente. 2006. "The Sociology of Expectations in Science and Technology." *Technology Analysis & Strategic Management 18*(3–4): 285–298.

Bourdieu, Pierre. 1996. *The Social Critique of the Judgement of Taste*. Translated by Richard Nice. Cambridge, MA: Harvard University Press.

Cayley, David. 2005. *The Rivers North of the Future: The Testament of Ivan Illich*. Toronto: House of Anansi Press.

Certeau, Michel de. 1990 [1980]. *L'invention du quotidien, tome 1: Arts de faire*. Paris: Gallimard.

Dalgalarrondo, Sébastien, and Tristan Fournier. 2019a. "Optimization as Morality, or the Routes of the Self." Translated by Rachel Gomme. *Ethnologie Française 176*(4): 639–651.

Dalgalarrondo, Sébastien, and Tristan Fournier. 2019a. "Un transhumanisme 'à mains nues.' Sociologie de la promesse du jeûne." *Revue d'Anthropologie des Connaissances 13*(2): 559–584.

Dalgalarrondo, Sébastien, and Boris Hauray. 2015. "Les économies de la promesse anti-âge. Le cas de la DHEA." *Sciences Sociales et Santé 33*(2): 5–30.

Deleuze, Gilles, and Félix Guattari. 2005. *A Thousand Plateaus: Capitalism and Schizophrenia*. Translated by Brian Massumi. Minneapolis, MN: The University of Minnesota Press.

Dubuisson-Quellier, Sophie, ed. 2016. *Gouverner les conduites*. Paris: Presses de Sciences Po.

Dupuy, Anne. 2013. *Plaisirs alimentaires. Socialisation des enfants et des adolescents*. Rennes: Presses Universitaires de Rennes/Presses Universitaires François Rabelais.

Fournier, Tristan. 2014. "Face à l'injonction diététique, un 'relativisme nutritionnel' en France." *Sociologies, Premiers textes*. http://sociologies.revues.org/4628.

Fournier, Tristan, and Olivier Lepiller. 2019. "Se nourrir de promesses. Enjeux et critiques de l'introduction de deux innovations dans le domaine alimentaire: Test nutri-génétique et viande in vitro." *Socio 12*: 73–95.

Gooldin, Sigal. 2003. "Fasting Women, Living Skeletons and Hunger Artists: Spectacles of Body and Miracles at the Turn of a Century." *Body & Society* 9(2): 27–53.

Griffith, Ruth Marie. 2000. "Apostles of Abstinence: Fasting and Masculinity during the Progressive Era." *American Quarterly 52*(4): 599–638.

Gueguen, Juliette, Isabelle Dufaure, Caroline Barry, and Bruno Falissard. 2014. *Évaluation de l'efficacité de la pratique du jeûne comme pratique à visée préventive ou thérapeutique*. Paris: Éditions Inserm.

Hadot, Pierre. 1995. *Philosophy as a Way of Life*. Edited by Arnold Davidson and translated by Michael Chase. Oxford: Blackwell Publishers.

Hervieu-Léger, Danièle. 2017. *Le temps des moines. Clôture et hospitalité*. Paris: Presses universitaires de France.

Jasanoff, Sheila, and Sang-Hyun Kim. 2015. *Dreamscapes of Modernity: Sociotechnical Imaginaries and the Fabrication of Power*. Chicago and London: University of Chicago Press.

Lefebvre, Henri. 1997. *Critique de la vie quotidienne, tome 1. Introduction*. Paris: L'Arche.

Lepiller, Olivier. 2013. "La bouffe industrielle à la casserole. Les effets de la critique de l'alimentation industrielle par la nutrition et les diététiques alternatives (1965–1985)." In *La juste mesure: une sociologie historique des normes* alimentaires. Edited by Thomas Depecker, Anne Lhuissier, and Aurélie Maurice, 115–144. Rennes: Presses Universitaires de Rennes/Presses Universitaires François Rabelais.

Lepiller, Olivier, and Yount-André Chelsie. 2020. "Quand la défiance frémit. Une assiette qui se politise à petit feu?" In *Quand l'alimentation se fait politique(s)*. Edited by Éve Fouilleux and Laura Michel, 49–64. Rennes: Presses Universitaires de Rennes.

Lestrade de, Thierry. 2015. *Le jeûne, une nouvelle thérapie?* Paris: Éditions La Découverte.

Macfadden, Bernarr, and Felix Oswald. 1900. *Fasting, Hydropathy and Exercise. Nature's Wonderful Remedies for the Cure of All Chronic and Acute Diseases*. New York: The Physical Culture Publishing Co.

Mathé, Thierry, Gabriel Tavoularis, and Thomas Pilorin. 2009. "La gastronomie s'inscrit dans la continuité du modèle alimentaire français." *Cahier de Recherche du CREDOC 267*.

Ong, Aihwa, and Stephen J. Collier, eds. 2008. *Global Assemblages: Technology, Politics, and Ethics as Anthropological Problems*. Malden, MA: John Wiley & Sons.

Poulain, Jean-Pierre. 2009. *Sociologie de l'obésité*. Paris: Presses Universitaires de France.

Poulain, Jean-Pierre, Romain Guignard, Claude Michaud, and Hélène Escalon. 2010. "Les repas: Distribution journalière, structure, lieux et convivialité." In *Baromètre Santé Nutrition 2008*. Edited by Hélène Escalon, Claire Bossard, and François Beck, 187–213. Saint-Denis: Inpes.

Pruvost, Geneviève. 2021. *Quotidien politique: Féminisme, écologie, subsistance*. Paris: La Découverte.

Rapport NACRe. 2017. *Jeûne, régimes restrictifs et cancer: Revue systématique des données scientifiques et analyse socio-anthropologique sur la place du jeûne en France*. Paris: Réseau Nutrition Activité physique Cancer Recherche.

Scott, James C. 2008. *Domination and the Arts of Resistance: Hidden Transcripts*. New Haven, CT: Yale University Press.

Scrinis, Gyorgy. 2013. *Nutritionism. The Science and Politics of Dietary Advice*. New York: Columbia University Press.

Shelton, Herbert Macgolfin. 1934. *The Hygienic System: Fasting and Sun Bathing*. San Antonio, TX: Dr. Shelton's Health School.

Sloterdijk, Peter. 2014. *You Must Change Your Life*. Translated by Wieland Hoban. Cambridge: Polity Press.

Thoreau, Henry David. 1854. *Walden, or Life in the Woods*. Boston: Ticknor and Fields.

Wolff, Francis. 2017. *Trois utopies contemporaines*. Paris: Fayard.

Zask, Joëlle. 2010. "Sortir hors de soi." In *La voix et la vertu. Variétés du perfectionnisme moral*. Edited by Sandra Laugier, 397–416. Paris: Presses Universitaires de France.

6 The self, the other and the world

The issues of today's fasting practice

Ophélie Bidet

The primitivist process of self-optimization is the best known and best documented dimension of the practice of fasting (Longo and Mattson 2014; de Lestrade 2015; de Toledo et al. 2019). Yet the political dimension of fasting – its cousin protest form of hunger strike apart (Siméant 2009) – is gradually gaining visibility, as presented by Sebastien Dalgalarrondo and Tristan Fournier in their chapter. Through this commentary, I would like to stress on the fact that fasting, more than a technique for optimizing oneself, can also be seen as a technique for optimizing oneself *in the world*, thus politicizing one's daily life. Indeed, fasting can link the faster to the surrounding world, enable them to question this relation and, as the case may be, to sustain or to initiate its transformation. The goal of such a transformation would tend toward a healthier way of being in the world, one that would be more respectful of nature and of other human beings. It thus reveals the possible merging of fasting into a *(re)wilding* perspective for the fasting person. This rewilding process seems to follow four paths that will be presented throughout this commentary: eating, consuming, ecology and a sense of solidarity.

The fasting practice which will be presented hereafter is that of more than a 1,000 fasters from France and French-speaking Switzerland,[1] aged from 20 to over 80 years old (the average age being 58). Eighty-two percent of them define themselves as women and 18% as men; 59% have a higher education degree; and 86% have already fasted for several days in a row, among whom 18% fasted for the first time more than ten years ago. But for nearly half of them, the first fast took place in the last three years.[2] Although with no absolute certainty, it can be estimated that this sample reflects the practice of people who fast outside of institutional religious settings in Europe.

1 The data are drawn from a PhD thesis conducted at the University of Neuchâtel, Switzerland, on the evolution of fasting outside a religious context and its inclusion into lifestyles. It includes in particular 35 semi-structured interviews and a questionnaire completed in full by 1,134 fasters. All data were collected in 2020 and 2021.

2 It would correspond to the 2018–2020 period.

DOI: 10.4324/9781003243496-12

The hunger that liberates

The link between fasting and eating may seem obvious, as it is first and foremost an eating practice (albeit a non-eating one). However, the questionnaire survey shows that 80% of the fasters have introduced changes in the way they eat after they experienced fasting, and 47% of them declare that their principal motivation to fast is to balance their relationship with food and to change their eating habits. These changes are usually implemented as soon as the fast is over. Some will last over time, others will fade away little by little, and a new fast can then be seen as a "reminder"[3] to help sustain these changes over time. These fasters speak of fasting as a "rest," a "renewal" or a "reset" allowing them to be hyper-reflexive about their relationship with food . . . via the absence of food.

First, experiencing fasting can allow a "liberation" from certain social injunctions or conventions related to food, for example, having three meals a day and at specific hours or the importance of breakfast. Thus, fasting creates an optimal opportunity to become conscious of how the individual is embedded in a feeding structure with its dogmas, but also its routines, and to deconstruct and possibly distance oneself from them.

> You really dissociate the desire for food and the hunger. . . . And I think it's great . . . to become conscious of this difference, because in everyday life, you don't ask yourself "Am I hungry?" The food is there and you help yourself. And I find it interesting too, what happens in the mechanism of food.
>
> (Tiphanie, 41-year-old woman, fasted for the first time five years prior)[4]

> [Fasting] is just detaching yourself from the daily routine that may have been imposed on you. . . . And at the end of the fast, to tell yourself: "Ok, in fact my thing is to have no breakfast. I don't like breakfast, I've always had breakfast because I've been told that it's the most important meal of the day, and actually, when I wake up in the morning, I'm not hungry. So I'm going to eat at lunch time when I'm really hungry."
>
> (Judith, 28-year-old woman, first fast: four years prior)

A survival technique for wild times

The "real" hunger of the body (not the desire dictated by the mind or by social norms) is thus put forward as the basis of the feeding rhythm. Fasting can also allow a "liberation"[5] from fears linked to hunger or the fear of dying of starvation.

3 The words in inverted commas are from the fasters' answers to the questionnaire.
4 These are verbatim from interviews. The names have been anonymized.
5 The words in inverted commas are from the fasters' answers to the questionnaire.

Some people have noticed that fasting enables them to have a stronger resilience when faced with certain situations that may involve a (risk of) lack of food, and that they therefore feel more serene in the way they deal with uncertainty.

> [Fasting] has allowed me to be more detached. In particular, the famous outing when you leave and you absolutely have to have a snack in your pocket just in case. . . . Well, now I don't need to have a snack in my pocket. I say to myself: "Well, it's not the end of the world if I once have to wait three hours before I eat." Yes, there was a little detachment from food, from the "risk that you don't eat" aspect.
>
> (Marthe, woman in her late thirties, first fast around five years prior)

These aspects are perfectly illustrated by the feeling of some fasters in the face of the Covid-19 pandemic, and in particular in the face of the episodes of panic buying in the supermarkets during the first lockdown period.

> Fasting has changed my life. You feel stronger. . . . You realize that many things can happen to you. . . . You see, there, the lockdown, I said to myself: "Well, if we can't go shopping, I'll fast!" When there were the yellow vests protests[6] at one point they said they were going to block everything. Well, I thought: "Ok, I'm staying at home, I'm not eating."
>
> (Gabrielle, 52-year-old woman, first fast 15 years prior)

Knowing that you can survive (and already have) several days or several weeks without food is a game-changer, and "helps you with fears."

> Let's say there is an even bigger health catastrophe, I know I can fast for several weeks, so it's not a big deal at all.
>
> (Vivianne, woman in her early fifties, first fast 18 years prior)

Fasting can thus be considered as a survival technique, a natural and wild response of a body that "knows how," in the face of uncertainty, and even in the face of the rewilding of the world as in the case of the pandemic linked to the Covid-19 virus (Searle and Turnbull 2020; Hornborg 2021; Soga et al. 2021).

The faster as a consumer

In addition, the well-being felt during and after this experience of "less" in feeding can allow the faster to become aware of the "too much" in their feeding practices:

6 The yellow vests protests – known as "gilets jaunes" in French – refer to protests and blockages that took place for weeks in multiple places around France after a rise in gas prices in late fall 2018, leading to violent clashes with the police. They are named after the hi-vis jackets worn by the demonstrators as a rallying sign. See, for example: www.bbc.co.uk/news/resources/idt-sh/yellow_vests.

too much food, too much fat, too much sugar, too fast, food that is too processed or too polluted. . . . People who fast can thus develop a desire to eat better, a quest that applies to both quantity and quality. Many report it as a strong impulse that comes from their body, a body which they have learnt to listen to better.

> At the end of a fast, I have absolutely no desire for this kind of dish [pizza], none whatsoever! My body wants to eat a good salad, a good ripe tomato, eat fruit, things like that. And I've noticed that I don't feel like eating things . . . nothing prefabricated, and canned stuff, just the smell of it makes me sick.
>
> (François, 60 years old man, first fast 26 years prior)

On the one hand, fasters choose to consume less meat, dairy products, salt, sugar and alcohol, to increase their consumption of vegetables, fruits, whole foods, raw or steamed, and even to implement intermittent fasting in their daily routine. On the other hand, it is a matter of decreasing the consumption of industrial, refined, and processed products, increasing the consumption of organic, local, raw products, from short circuits. . . . In short, it is about adopting new consumption habits, by returning to the simple, the natural and the wild, by following the body as a guide.

The reflexivity on food practices that fasting allows is thus extended to consumption practices in general, on a larger scale, as illustrated by the fact that 36% of fasters declare that one of their motivations is to reflect on their consumption habits. Many talk about fasting as an experience of "joyful sobriety" that "makes one think about the material" and makes one want "more from less." Seventeen percent of fasters expressly consider their fasting as a manifestation of their disagreement with the consumer society.

> It is absolutely necessary to distance ourselves from this consumer society. Fasting is a very good way to do that. Because afterwards we can distance ourselves from many other things, too, that we think we need, but don't.
>
> (Irene, 75-year-old woman, first fast 20 years prior)

> And what is surprising is that, when you go shopping at the end of the fast to buy things for refeeding, if you go to a big shopping center, you look at the shelves and you say: "We don't need 90% of these." We don't need 90% of all these processed products, stuff, etc.
>
> (Diane, 68-year-old woman, first fast between 10 and 15 years prior)

Into the wild

This reflection on the modes of consumption, (over)production and the society of (over)consumption brought about by fasting comes to integrate ecological problems naturally, such as pollution, the finitude of resources and global warming.

The ecological coloring of this reflection is also nourished by the natural dimension of the practice of fasting. Indeed, the relationship between fasting and nature is fully confirmed by the contemporary practices of fasters. In particular, for many fasters, it is more difficult to fast in an urban environment, "more aggressive for the senses." They prefer the calm and natural setting of the countryside, with "less noise, less pollution" and far from everyday life.

For 74% of them, this also means walking in the middle of nature during their fasting – which is also very developed in fasting centers with the concept of "fasting and hiking"[7] (Cohen, Bellenchombre, and Féliu 2019; Dalgalarrondo and Fournier 2019, 2020). Rewilding in the sense of reconnecting to nature, "to the elements," to "Pachamama" – to use the words of the fasters – is thus an integral part of the contemporary experience of fasting, and even of the motivations for fasting. The fasting person thus finds themselves, through their practice, repositioned within the natural world that they inhabit (Mathevet, Thompson, and Bonnin 2012), not only as an eater and a consumer but also as a living being that is part of a whole. A whole . . . with others.

When the gaze shifts outward

Although it is experienced in comfortable and "extremely artificial" conditions, some fasters share that the experience of fasting – and hunger – has made them aware of the daily lives of a large number of people on the planet, who are in precarious and insecure food situations. The dimension of solidarity in fasting thus appears as follows, as Boudier relates in her experience of Ramadan (1999):

> When I feel hunger . . . in those moments, I say to myself "there are so many women raising their children who don't have enough to eat at all, and there are so many people on earth who go through this all the time, all the time, all the time!"
>
> (Irene)

> It was during a fast that I realized that there were people elsewhere on the planet who were suffering from hunger. I have known this for a long time, I have done university studies and so on. I know that. But it was really through fasting that I became aware of human suffering.
>
> (Klara, woman in her mid-forties, first fast 13 years prior)

It is the chosen dimension of the fast and its limited duration agreed upon in advance that opposes to the uncertainty of hunger suffered by precarious populations.

> During this week [of fasting], I'm convinced that I'm fasting by choice, but there are millions of people who don't fast by choice and who have to fight all

7 See, for example, the website of the French Federation of Fasting and Hiking: www.ffjr.com/.

> day to get something to eat at night. And [fasting] makes you aware of that too.
>
> (Johannie, woman in her seventies, first fast 14 years prior)

> It means realizing how lucky we are to have access to food in abundance . . . and to think that we really live in the luxury of being able to decide when we want to eat or not, and it's true that this is not the case for everyone.
>
> (Judith)

As Irene explained when interviewed, fasting can "connect [the fasting person] with these people," whether they are down the street or at the other end of the planet. Solidarity with the destitute, the abandoned and the excluded is an active part of the intentions of 10% of fasters when they fast. This dimension of solidarity is, for example, the cornerstone of the ecumenical fasting groups "Fasting together"[8] offered in Switzerland, or of the association "Calories for Life,"[9] also born in Switzerland. The fasting person is invited to donate the money that they have not spent on meals during the fast to associations or NGOs that fight world hunger. The main victims of the effects of climate change are the populations vulnerable in their access to food and resources (Laurent 2009; World Meteorological Organization 2021).

Finally, all these dimensions merge into the ecological movement and thinking: a call for a change in food and consumption practices, to protect nature, for solidarity toward precarious populations but also toward future generations, a better management of resources. . . . Consequently, these dimensions participate in the will to implement or reinforce more environmentally friendly practices (organic, local, vegetarianism, green mobility, secondhand objects, etc.) following a fast, or to engage more actively in them. They thus support a more sustainable *rewilding* of the faster. Twenty-four percent of the fasters declare that they fast to live a militant approach of sobriety out of respect for nature and future generations. This approach is perfectly modeled in the "Fast for Climate" initiative. Born before COP21 in 2015 and carried by the Philippine delegate to the UN Yeb Sano, this invitation to fast is meant to be a manifestation of the desire to break with current behaviors to act against global warming and demand political action. The call to fast for the climate summarizes the intertwining of these different dimensions in the practice of fasting very well, which is presented as an ideal tool to "reflect on our way of inhabiting the world" in its entirety:

> Fasting is about taking care. Taking care of ourselves by offering our body and mind a different experience. Taking care of others by showing our ability to limit ourselves for a better distribution of resources. Taking care of what

8 https://voir-et-agir.ch/pour-les-paroisses/groupes-de-jeune/.
9 https://descaloriespourlavie.ch/.

some call nature, others creation, by showing ourselves capable of resisting the greed encouraged by our consumer societies.

(Excerpt from the call to fast for the climate on the occasion of COP26)[10]

In conclusion, fasting can offer a space for reflection for the fasting person on their relationship to the world around them, while at the same time presenting itself as a resource to support a response to several issues (notably relating to climate, (over)consumption and solidarity) that could be raised in this way. In particular, it would enable people to reclaim skills in the areas of food and cooking or in listening to their bodies – skills that contemporary lifestyles often neglected. The practice of fasting supports the (re)discovery of the "confidence that [one's] body has a certain intelligence," a certain autonomy, a *wild* potential that can prove to be resourceful in situations such as the Covid-19 pandemic.

References

Boudier, Séverine. 1999. "Le Ramadan serait-il contagieux?" *Autres Temps. Cahiers d'Éthique Sociale et Politique* 62: 21–25. https://doi.org/10.3406/chris.1999.2118

Cohen, Patrice, Laura Bellenchombre, and François Féliu. 2019. "Jeûner en France. Généalogie d'une pratique contestataire et contestée." *Revue des Sciences Sociales* 61(1): 100–109. https://doi.org/10.4000/revss.3701

Dalgalarrondo, Sébastien, and Tristan Fournier. 2019. "Un transhumanisme 'à mains nues': Sociologie de la promesse du jeûne." *Revue d'Anthropologie des Connaissances* 13(2): 559–584. https://doi.org/10.3917/rac.043.0559

Dalgalarrondo, Sébastien, and Tristan Fournier. 2020. *L'utopie sauvage*. Paris: Les Arènes.

de Lestrade, Thierry. 2015. *Le jeûne, une nouvelle thérapie?* Paris: La Découverte.

de Toledo, Françoise Wilhelmi, Franziska Grundler, Audrey Bergouignan, Stefan Drinda, and Andreas Michalsen. 2019. "Safety, Health Improvement and Well-being during a 4 to 21-day Fasting Period in an Observational Study Including 1422 Subjects." *PloS One* 14(1). https://doi.org/10.1371/journal.pone.0209353

Hornborg, Alf. 2021. "Beyond the Image of COVID-19 as Nature's Revenge: Understanding Globalized Capitalism Through an Epidemiology of Money." *Sustainability* 13(9): 5009. https://doi.org/10.3390/su13095009

Laurent, Éloi. 2009. "Écologie et inégalités." *Revue de l'OFCE* 109: 33–57. https://doi.org/10.3917/reof.109.0033

Longo, Valter D., and Mark P. Mattson. 2014. "Fasting: Molecular Mechanisms and Clinical Applications." *Cell Metabolism* 19(2): 181–192. https://doi.org/10.1016/j.cmet.2013.12.008

Mathevet, Raphaël, John Thompson, and Marie Bonnin. 2012. "La solidarité écologique: prémices d'une pensée écologique pour le xxie siècle?" *Écologie & Politique* 44: 127–138. https://doi.org/10.3917/ecopo.044.0127

Searle, Adam, and Jonathon Turnbull. 2020. "Resurgent Natures? More-Than-Human Perspectives on COVID-19." *Dialogues in Human Geography* 10(2): 291–295. https://doi.org/10.1177/2043820620933859

10 https://jeunepourleclimat.net/appel/.

Siméant, Johanna. 2009. *La grève de la faim*. Paris: Les Presses de Science Po.
Soga, Masashi, Maldwyn J. Evans, Daniel T.C. Cox, and Kevin J. Gaston. 2021. "Impacts of the COVID-19 Pandemic on Human – Nature Interactions: Pathways, Evidence and Implications." *People and Nature* 3: 518–527. https://doi.org/10.1002/pan3.10201
World Meteorological Organisation. 2021. *State of the Global Climate 2021: WMO Provisional Report*. 1264. https://library.wmo.int/doc_num.php?explnum_id=10859

Part 3

The wild at the interstices

A route to empowerment?

Conversation 4. Rummaging within the city: Wild plants and edible rubbish

7 Foraging plants within the urban margins

On the possibilities of living with nature in the greater Paris

Flaminia Paddeu and Fabien Roussel

Introduction

Gathering plants usually takes place above all in rural, forest and mountain territories, whose ecologies and social economies seem more favorable (see Bromberger and Lenclud 1982 for the issue of *Etudes Rurales* devoted to hunting and gathering in rural areas). Yet discreet and often-neglected, urban foraging is nevertheless considered a global phenomenon (Paddeu 2019). Urban foraging is the practice of harvesting or gathering raw biological resources (plants and fungi) within urban settings, primarily for direct food or medical consumption, decoration, crafts, ceremony, barter or small-scale sale[1] (Shackleton et al. 2017). It can include wild or domesticated species, in managed or unmanaged spaces. It takes place in varied settings of urban, suburban and peri-urban landscapes, including common, state and privately owned lands. In cities of the Global North, it is practiced more or less openly by the rich and the poor, the employed and the unemployed, migrants and long-term residents, whose motives and practices vary in time and place. Generally bereft of monetary exchange, it relies on specific skills to locate, gather and sometimes process and distribute the resources. Usually prohibited by local urban governments, foraging mostly tends to be simply overlooked.

We use the practice of gathering plants as our foremost object to grasp relationships with urban spaces and ecologies, especially in the margins. Foraging has recently gained momentum within cities such as Seattle, Baltimore, New York, Philadelphia or Syracuse (McLain et al. 2014; Plieninger et al. 2015; Synk et al. 2017), Stockholm, Berlin or Edinburgh (von Hoffen and Säumel 2014), and in Finnish and Japanese towns (Terada et al. 2010; Kangas and Markkanen 2001). Yet little has been researched on foraging in the Greater Paris area (except for Legrand 2017). Composed of 131 municipalities and more than seven million inhabitants,[2] this territory has been undergoing a metropolization process since the 1980s (Gilli 2014). It has resulted in economic, social and spatial transformation, linked in particular to deindustrialization and tertiarization, increasing

1 In this chapter, we specifically focus on plants and food uses, even if we keep other kinds of uses in our scope, as they are sometimes closely intertwined.
2 According to metropolegrandparis.fr.

DOI: 10.4324/9781003243496-15

fragmentation between varied land uses and segregation between different populations. It is also a major factor in the growing artificialization of the land, as the areas around the ring road and in the inner as well of outer suburbs, especially around the stations of the Greater Paris rail network, are considered ideal locations for new urban projects. In this changing geography, recognized urban parks and forests are dedicated to recreation and biodiversity conservation. Yet wastelands and vacant lots are increasingly considered land reserves and coveted spaces for urbanization, despite green and ecological urban policies of open spaces (Flegeau and Roussel 2021). In this context, margins are not only defined as the space peripheral to the urban center, but they are also an alternative place to the dominant system: either a space of relegation of some practices, population and plants, or a space of freedom and invention of new ways of interactions (Grésillon, Alexandre, and Sajaloli 2016). Thus, marginality can also emerge in the heart of the city, whether it is related to the action of individuals or of plants, but also the alliances that they form.

Drawing upon the literature and our research, we defend a social geography and urban political ecology approach that focuses on the intersection of foraging, urban botany and nature management in a dense metropolitan context. Thus, the metropolis appears as a new "botanical field" of investigation (Gandy and Jasper 2020), intertwining ecological dynamics of plants in a disturbed environment, urbanization process and urban policies of nature. In the background, it questions the rewilding possibilities in the city and the role that foraging plants plays in that direction. In France, law n° 2014–110, known as the "loi Labbé" of 6 February 2014, marked the abandonment of the use of pesticides in the management of public urban green spaces. This law, intended to preserve biodiversity, has been a real boon for foraging, by reducing the risk of poisoning on the one hand, and allowing a number of synurbic species prized by foragers (Francis and Chadwick 2012) to gain ground in the city on the other hand.

In this chapter, we wish to share our initial questions and first results. We make the hypothesis that foraging is a practice allowing to analyze entangled material and cultural interactions with plants in the city and, moreover, with urban ecologies and spaces. First, urban foraging enables us to better grasp relationships to the metropolis, and especially its spatial margins. We made the assumption that foraging happened in marginalized places, such as wastelands, vacant lots, edges, abandoned orchards or interstitial spaces. Yet we discovered that foraging takes place in a continuum of spaces that are more or less integrated or marginalized regarding metropolitan areas. Second, focusing on gathering plants in the city allows us to better understand relationships to ecology and the living world within a large metropolis. We started with the idea that foraging was embedded within ecological mindsets and habits, allowing urbanites to reconnect with "nature" through the practice of botanical identification and consumption. We realized that there is an array of gatherers' relationships with plants and their uses, according to their socio-economic status and their spatial trajectory, widening what we traditionally grasp as ecological. Third, the regulation of foraging is a way to analyze the relationships of urban policies to ecology, and what kinds of interactions with plants are mainly encouraged or deterred. We made the hypothesis that foraging was an overlooked practice by urban managers. As a matter of fact, it is rather considered a marginal yet subversive practice for urban policies of nature management, hardly compatible with the dominant biodiversity paradigm.

Our ongoing geographical survey of foraging in the Greater Paris area is based on the collection of interviews and *in situ* observations. We used a qualitative social sciences methodology, combining 21 semi-structured interviews and 23 informal conversations with individual gatherers, leaders of foraging organizations and urban nature managers, and participatory observations on gathering sites. We focused our attention on active urban gatherers, who dedicate a specific timespan and interest in foraging within metropolitan spaces, contrary to recreational gathering, regarding people who forage occasionally in peri-urban forests for common species (chestnuts, mushrooms, etc.).[3] We also undertook an ethnobotanical analysis of plants in order to highlight the species that are targeted, those that are rejected, and also the representations and role attributed to each species.

Following our three hypotheses, the first part focuses on the spatial dimension of gathering, the second part explores the gatherers' relationships to plants and the third part discusses the confrontation to urban policies of nature management.

1. A continuum of places more or less integrated to the metropolis

1.1. *Diverse study areas within the Greater Paris*

The areas selected for the field surveys are located in the Paris metropolitan area, known as the Greater Paris. This sector was chosen because, on the one hand, there was a lack of knowledge on the subject of foraging. On the other hand, we both benefit from past research experiences in this area, whether citizen environmentalism and urban agriculture (see Paddeu 2021) or the socio-ecological dynamics of the Île-de-France' green belt (see Roussel and Alexandre 2021). The choice of study areas is based on the places covered by people, companies and institutions surveyed. Eventually, those places reflect the diversity of the Greater Paris area on both morphological (Figure 7.1) and socio-economic levels (Figure 7.2).

3 Recreational gatherers refer to a more rural way of interacting with plants and spaces: they mostly target public forests and do not engage with a critical approach of the city or the urban environment. Yet we believe that the playful dimension is common to different types of gathering and refers to the pleasure of foraging. For instance, Nicole, a 50-year-old psychologist apple picker living on the periphery of the metropolis acknowledges that she "*doesn't know anything about apples*" but spending times in abandoned orchards gives her "*the impression of being in August Renoir's time.*" When asked about her motivations, she says: "*So the beauty of the landscape, not letting fruit rot, and then I find it very pleasant, it must be my squirrel side.*"

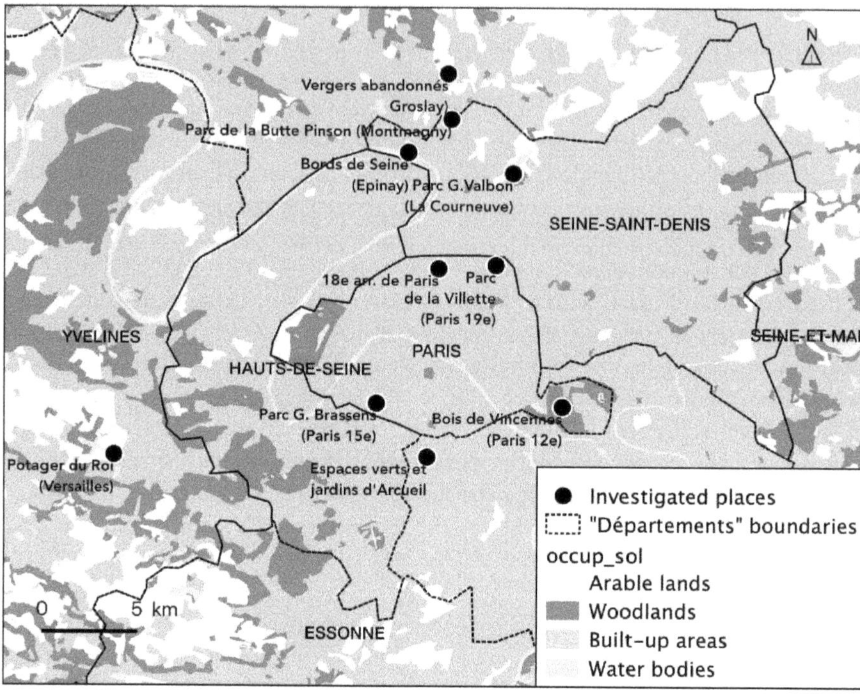

Figure 7.1 Study areas and land cover
Authors: Roussel and Paddeu

From a morphological standpoint (Figure 7.1), the study areas are sometimes located in the heart of the conurbation, in densely built-up areas (the 15th, 18th and 19th *arrondissements* of Paris, the municipality of Arcueil), and sometimes on the inner fringes of the conurbation (La Courneuve and Epinay) or even on the outer fringes, close to the rural ring (Versailles, Montmagny and Groslay). This reflects the diversity of green spaces: densely built-up and populated areas, neglected interstices, vast urban parks and, further away from the heart of the city, agricultural wastelands or public forests.

From a socio-economic standpoint, the selected areas are also representative of the contrasted situations in the metropolitan area. In the southwest, the metropolis is characterized by wealthy populations, in the north and east by more disadvantaged ones. This is reflected in the median income per household and per municipality (Figure 7.2). The areas surveyed in La Courneuve, Epinay Montmagny and also the 18th and 19th *arrondissements* of Paris correspond to less privileged situations (fewer than 22,000 euros per year per household), whereas Versailles and the 12th and 15th *arrondissements* of Paris are more privileged (more than 25,000 euros per year per household).

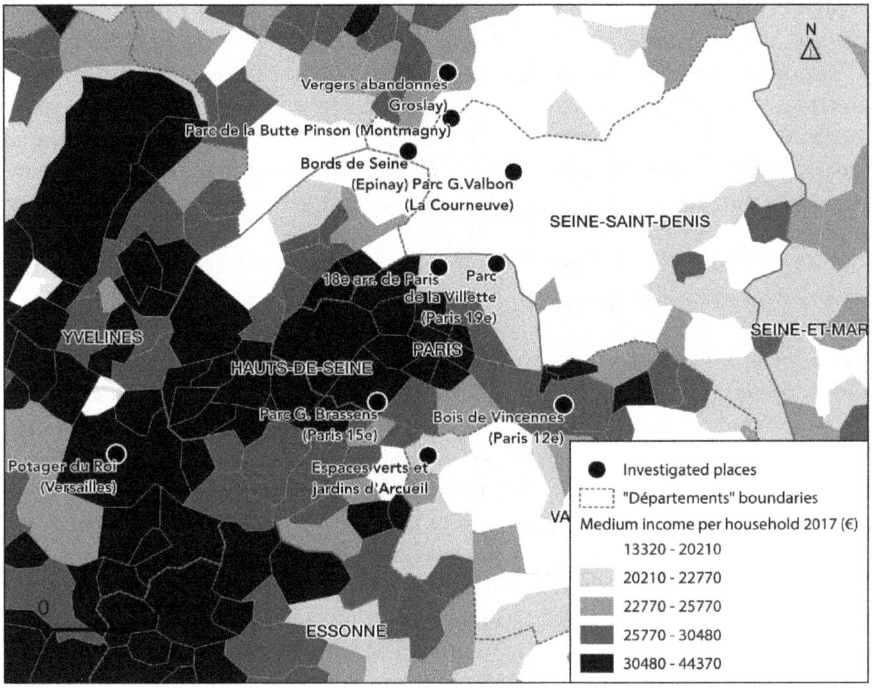

Figure 7.2 Study areas and socio-economic contexts

Authors: Roussel and Paddeu

1.2. The micro-territorialities of foraging

The geography of urban foraging spaces can be analyzed at two scales. First is analyzed at the scale of the metropolis, in which various categories of green spaces are concerned (Figure 7.1). If gathering plants is a diffuse practice, it is widespread in all types of urban and peri-urban green spaces. First, it takes place within the forests, such as the Bois de Vincennes, but also within the peri-urban state forests such as Montmorency. A second type of spaces is the urban parks and gardens, such as Parc de la Villette in the 19th *arrondissement* or Parc Georges Brassens in the 15th *arrondissement*. In both of these categories, some organizations set up gathering events, such as a botanical outing we attended in Parc Georges Valbon in La Courneuve. Third, we discovered that abandoned orchards are also targeted, such as the apple or pear orchards located in Groslay. They are a vestige of an activity that has largely declined and thus given way to abandoned rewilded areas. They are sometimes valued for their biodiversity, but are more often concerned with informal practices such as waste deposit or campsites for the Rom community. Finally, urban gatherers also forage in urban interstices, hollows and edges in the most mineral sectors of the city. A few gatherers collect plants in flower beds located near the street, at the root of a street tree, in the cracks of the pavement; whereas others consider these micro-spaces as dirty and unfit for gathering.

Second, at the scale of the green spaces, there is a specific geography of foraging that targets margins at a micro-local scale. The spatial edges and the derelict spaces specifically attract urban gatherers. New biodiversity practices, such as differentiated management of green spaces, create a park zoning between parts that are taken care of, mowed, cut, flowered and parts where weeds are accepted, meadows can grow or undergrowth spaces are left alone. For instance, some spaces of urban parks are left wild, often on the edges, such as the embankment of the Parc Georges Brassens that adjoins an abandoned city railroad. These practices enable micro-territories of foraging to appear inside the green spaces. Two reasons seem to explain these patterns. First, it gives urban foragers access to disregarded species in urban contexts such as nettles (*Urtica dioica*), dandelions (*Taraxacum officinal*) or blackberries (*Rubus fruticosus*), some of the most foraged plants among the 128 species encountered in our survey. Second, it allows foragers to remain discrete, as dense vegetation can act as a screen to protect them from the sight of urban gardeners and foragers.

Sometimes, gatherers unexpectedly target green areas that are managed for landscape and aesthetic purposes, such as flower beds and other ornate spaces. Thus, they divert the primary function through their personal practice creating tensions with park managers and rangers such as in Parc G. Valbon or Parc G. Brassens. In these cases, gathering interferes with the traditional management of green spaces as a decoration. It produces a form of low-key disruption by interfering with the established framework. As these practices of collecting ornamental plants concern various types of gatherers, park rangers tend to specifically associate these damaging practices with immigrant populations from Asian origin, although they are practiced by a wider range of population.

There may also be health risks associated with urban gathering that limit its potentialities and changes spatial patterns of gathering. In addition to potential misidentification of species, people consuming materials foraged in urban settings may be exposed to chemical contaminants (von Hoffen and Säumel 2014; Synk et al. 2017; Weeks et al. 2006), although very few cases of sickness due to urban foraging have been reported in the literature (Poe et al. 2013). Most gatherers actually adapt their practices to avoid health hazards. Foragers stay away from sites with contaminated soils, exposed to traffic pollution or soiled: "*For example, there is a street where purslane grows, but purslane likes to be trampled on, so I don't feel like picking it, because it is also potentially sprayed by dogs, on a pavement where cars are parked, not far from the exhaust pipes*" (Anne-Marie).

1.3. A mobile practice adapting to a constantly changing urban environment

The exploration of routes and habits of urban collecting, that is, "patterns of gathering" (Shackleton et al. 2017), reveals some aspects of urban transformations.

These transformations reconfigure the location, size, distribution, nature, management of, and access to the ecologies and green spaces that support gathering. Densification processes withdraw, fragment and deteriorate available places for foraging. Urban sprawl takes on a patchy form, resulting from the combination of zoning, land conservation and clustering policies. Hence, foragers seeking a diverse and substantial quantity of resources may need to cross a broader landscape to escape densification (Hurley et al. 2015; Synk et al. 2017). In the Greater Paris area, open spaces have been under great pressure from metropolitan or even international projects, such as the Greater Paris public transport network that is currently under construction. More recently, the developments planned for the Paris Olympic Games in 2024 are further eroding the green spaces of already dense areas.[4]

Conversely, through the production of wastelands, vacant lots and unbuilt areas, land abandonment is considered to provide gathering opportunities within urban areas. In almost all northern metropolises, and specifically in shrinking cities such as Detroit (Hurley et al. 2015; Millington 2013; Paddeu 2016), land markets and policies create land vacancy, providing transient opportunities for species to spur in abandoned areas and for gatherers to access edible species.

In this context, the gatherers we met show a great capacity for adaptation: most of them alternate between picking locations close to their homes within the urban area and less dense areas further away, or even go to rural locations. The practice is flexible and constantly reconfigures the places where people gather. Abandoned spaces certainly have an important role to play, but gatherers remain opportunistic. As a result, it is not so much the marginality of the spaces that defines the possibilities for gathering as the ability of the gatherers to move around. Foraging is first and foremost an activity of movement according to opportunities, as Nicole tells us about her peri-urban apple picking:

> *And then in Groslay, I had a lot of scramping zones, but they changed every year or two. That is to say, I would go by mountain bike, and I would say to myself, I'm going to come back to get my apples there, but I would come back the following year and I couldn't go anymore, or there were no more apples because there were too many brambles and it had become too dark.*

In this respect, it is opposed to the tradition of zoning and the compartmentalization of space in classical urbanism. It therefore raises the question of a right to wander (Vanuxem 2021).

4 See, for instance, the case of "Jardins des Vertus" in Aubervilliers: A.-R. Kokabi, "Sommés de partir, les jardiniers d'Aubervilliers résistent au Grand Paris," *Reporterre*, 16 juillet 2021, https://reporterre.net/Sommes-de-partir-les-jardiniers-d-Aubervilliers-resistent-au-Grand-Paris.

2. An array of gatherers' relationships with plants

2.1. *The city as a "planet garden"*

For the last 20 years, cities of the Global North have witnessed a revitalized interest for ordinary plant biodiversity, which values the flora of urban wastelands or the "weeds" that grow in the cracks of the asphalt (Machon 2010; Muratet et al. 2017). Foragers' attention to plants has to be considered within this context, but it also extends it as foragers target any plant – whether they belong to local habitats or come from abroad – as long as it is considered useful. In this sense, the plant ecology of foragers is closer to Gilles Clément's "planet garden" (Clément 2006) than biodiversity conservation policies that sort out plants depending on their rarity or local origins.

The 44 interviews, observations and informal discussions enabled us to identify more than 120 plant species[5] – other researchers documented 486 species in Seattle (Poe et al. 2013), 170 taxa in Baltimore (Synk et al. 2017) and 91 species in Sarajevo (Redzić 2010). Among these, the most commonly mentioned (more than five occurrences) are Nettle (*U. dioica*), Burdock (*Arctium lappa*), Dandelion (*T. officinal*) and Plantain (*Plantago* sp.) for herbaceous plants and Bramble (*R. fruticosus*), Hawthorn (*Crataegus monogyna*) and Black Elder (*Sambucus nigra*) for woody plants. All these species are commonplace in the urban ecological context of the Greater Paris. They can be qualified as ruderal and most of them are considered as weeds or simply go unnoticed in urban contexts. More generally, foraging challenges the well-established categories of desired and unwanted plants. Other criteria are at work here than aesthetic or ecological ones. They correspond to the medicinal or alimentary virtues of the plants, to the availability and abundance of the resource and also to the skills of the gatherers. The practice of foraging changes the way some depreciated species are considered. For instance, this is the case with species described as invasive in the context of ecological policies. The Black locust (*Robinia pseudoacacia*) is particularly appreciated for its flowers, which are cooked in beignets. More generally, harvesting blurs the ecological boundaries of species: it does not matter whether species come from other continents or have escaped from parks; it is their uses that are recognized. For instance, Mrs. Zheng takes advantage of the plantations of Japanese Sophora, Japanese Honeysuckle or Hibiscus by the city of Paris to cook according to traditions imported from China.

Preserving the resource is a source of concern for some gatherers. Some species, such as the Bear's garlic (*Allium ursinum*), are subject to spring rushes that endanger the plant's presence locally. Virginie says: "*We came back later to let the plants develop again and we went to another spot, in fact, because it was already a bit of a carnage.*" The concern to preserve the resource is constant among the pickers:

5 We count any plant that is mentioned by a gatherer during an interview or a field observation. Among the 128 mentioned, only ten are indicated for other purposes than consumption or medicine, especially toxicity Hemlock (*Conium maculatum*), Yew (*Taxus baccata*) or Common laburnum (*Laburnum anagyroides*).

all of them are self-reflexive about the quantities collected and their proven needs. Regarding Yarrow, which she uses to make infusions, Virginie says: "*I know the quantity I need for my year, it's the equivalent of a freezer bag this big, well filled with dry plants.*"

People usually forage year round, particularly in spring, summer and fall, which are active harvesting times owing to species availability, even if winter still offers gathering opportunities. It seems to be especially the case of specialists such as Mathias or Virginie who gather in all seasons, focusing on the different stages of development of plants (buds and flowers in spring, leaves and fruits in summer or autumn, roots in winter). Frequent outings both require and foster specific knowledge and relationships to places, goods and taxa that are collected. For the majority of professionals organizing foraging courses and some of the individual gatherers, we observed the acquisition of solid and precise knowledge in botany and ethnobotany. This is reflected in the knowledge of a wide and diversified range of species: some little known, such as the Galinsoga (*Galinsoga parviflora*), the Sweet woodruff (*Galium odoratum*) or Common tansy (*Tanacetum vulgare*). The attention to identification of plants is crucial, as risks of intoxication exist due to the ingestion of toxic plants, for instance Hemlock (*Conium maculatum*), because of its resemblance to the Wild carrot (*Daucus carota*): "*So afterwards, I don't take too many risks with umbelliferous plants because I can't distinguish between them, they're not my favorite plants, I get them mixed up a lot, and then there are risks*" (Anne-Marie).

The poor preparation or processing of a plant, the misuse of the consumable part of the plant, the overuse or incorrect dosage of a plant is also a source of concern – about Borage (*Borago officinalis*) Amandine explains: "*But if you eat it once in a while, it is no more dangerous than drinking a glass of wine*" – and so is the ignorance of the risks associated with a plant (e.g., plants that should not be consumed if you are pregnant, such as Mugwort (*Artemisia vulgaris*) or Tansy (*T. vulgare*).

Urban foragers thus often possess more or less sophisticated local ecological knowledge which they pass on through family and friends, ecological instruction, field guides and books (Poe et al. 2013). The acquisition of this knowledge and know-how were done through different channels: attending herbalism schools or university degrees; reading books and guides; the participation to training courses or workshops. It allows some of them to use the Latin name of the species or to master a specialized vocabulary. Thus, Amandine mentions "*pyrrolizidine alkaloids*" adding that "*if you eat a lot of them, a large quantity or often, it is dangerous.*" This knowledge is accompanied by know-how on gathering and processing plants. For example, from St. John's wort (*Hypericum maculatum*), Hélène extracts "*oil . . . which is used in case of burns.*" As for Amandine, she is currently learning paraculture, a traditional technique among the pygmies of Cameroon that consists of "*taking the part of the plant that will be used to feed them, and leaving the other part so that the plant starts again, i.e. the plant does not die.*" This knowledge combines Western scientific naturalist knowledge with non-Western vernacular knowledge.

For another part of the individual pickers and a minority of the professionals organizing courses, the relationship with picking is based on ordinary knowledge and know-how. This is the case of Nicole, an apple picker, who declares: "*I don't know anything about it. I know about flowers because I have a garden, but I know absolutely nothing about apples.*" She does not know the names of the apple varieties she picks but insists that they are "*old varieties.*" For these pickers, gathering focuses on common species of fruit (raspberries, blackberries, redcurrants, plums, apples, etc.), which are easily recognizable and identifiable by all, so that it can be done with people who are not trained in botany (e.g., women and children living in working-class neighborhoods for an association in Villetaneuse). This knowledge and know-how require more diffuse acquisition mechanisms: they are part of a common culture or have been passed on by relatives, such as grandparents, and are also supplemented by books such as nature guides.

The uses of plants are diverse but all are intended to be complementary to more conventional ways of consuming. Plants are not resources in their own right. We were not able to identify yet subsistence strategies as Kaoma and Shackleton (2014) observed in South Africa informal settlements for instance. In our survey, urban gathering is actually used to transform the composition and quality of individuals' diets. Food uses were the most frequently mentioned, whether they gave rise to processing, particularly for fruits (syrup, cream, preserves, jam, flour, juice, compote, lacto-fermentation, drinks and alcohol), or even to an elaborate gastronomic approach (velouté, tarts, gratin, mesclun, mousseline, quiche, pesto, flan, jelly, butter, salad, soup and cocktail). Some people simply eat the plants raw. Therapeutic uses were also mentioned in various forms (infusion, balm, oily macerate, decoction and spirit) and with objectives of well-being or alternative medicine: antispasmodic and analgesic virtues for Yarrow (*Achillea millefolium*), regulation of the cardiac rhythm for Hawthorn (*Crataegus monogina*). The issue of women's health came up: Hélène mentioned the fact that Nettle "*regulates the woman's cycle, stimulates the arrival of menstruation.*" The aesthetic dimension is sometimes evoked (bouquets, herbariums, drawing, plant dyeing, basketry and paper).

2.2. Common practices but distinct groups and contexts

Gatherers share some gestures, values and representations. Cooperation, gift giving and social reciprocity are also common among various types of gatherers, including marginalized people as well as integrated communities (Poe et al. 2013). However, urban gatherer demographics are particularly diverse in age, gender, income and ethnicity. Urban foraging has become popular among educated middle- and upper-class population sensitive to ecology, while it continues to be practiced by food-insecure individuals. In the Greater Paris as well as elsewhere (McLain 2014; Poe et al. 2013; Shackleton et al. 2017), foragers thus include indigenous and non-indigenous life-long residents, as well as domestic and international immigrants.

The urban foragers we surveyed are mostly white educated middle-class female and male, between 30 and 70 years old. Yet there is an array of profiles in terms of socio-economic status and integration. Foraging individuals are more or less economically stable and socially integrated, including individuals in precarious situations. On the one hand of the spectrum, Pierre-Emmanuel, a squatter in his thirties in the inner suburbs of Paris who cannot afford to pay for coffee in a café, struggles to make ends meet. On the other hand, Laure, a painting restorer in her late forties who regularly travels abroad or pays for expensive foraging training courses lives a comfortable metropolitan life. Most of them are college educated: Hélène followed a curriculum in urban planning, landscaping and botanical training, while Virginie and Amandine have diplomas in education or fine arts. Female gatherers are in the majority, and single women, divorced or widowed, are especially overrepresented. Yet male gatherers are far from absent, and the most famous herbalists tend to be men. Most of the urban gatherers we met have embedded foraging within a professional reconversion or a personal change in life. Some of them have rural origins or have family ties with rural areas, have lived there for a few years or plan to settle there.[6]

While certain gatherers continue to practice foraging as leisure, others engage in a professionalization process around their passion. After following training courses, they offer different types of social activities such as foraging outings, botanical arts and crafts, cooking classes or wild plant meals. They thus participate in the propagation of urban foraging in the context of collective formal or informal activities. If the territory is socially mixed, urban foraging social activities tend to appeal to middle- and upper-class educated white inhabitants.

In Bois de Vincennes, a metropolitan forest located in an upper- and middle-class area, the participants of botanical outings organized by a start-up are mostly white, female and highly educated. For instance, during one of these outings in July 2019, 11 participants out of 13 were women, between their 30s and 70s, working in the education, cooking and health sectors. When gathering activities occurs with underprivileged population, they are carried out by educated, white individuals, driven by social activist purposes. For instance, in Villetaneuse, a low-income municipality in the northern suburbs, two men in their thirties in their thirties, motivated by a local social project, set up a jam-cooking workshop with the gathered fruits of the forest. The participants are mainly North African housewives coming with their children to gather in a family-oriented local atmosphere.

Only a fraction of individual gatherers we met are recent or long-term immigrants, belonging to low-income or marginalized communities.[7] In the Greater Paris area, Chinese immigrants are especially common urban foragers, but we also encountered foragers from North Africa, the Middle East or Eastern Europe. Newly arrived foragers recognize species from their homelands and appreciate them differently. They are mostly employed in working-class jobs such as Samir, who works in construction, while others have informal jobs, such as Mrs. Zheng, a Chinese woman in her fifties who manages a "massage" salon that hosts a

6 Since the beginning of our investigation, at least four of the foragers we interviewed left the Greater Paris area to live in rural areas, whether before or during the Covid-19 pandemic.

7 Further investigation needs to be made to assess the quantitative importance of this population of gatherers, as our sample has not been built to be quantitatively representative of gathering practices in the Greater Paris area. Also, as particularly discreet and informal practices, they are more difficult to grasp through interviews and observations.

prostitution activity. They mostly practice foraging individually or with family, and are relatively disconnected to the associative and professional emerging sector of urban gathering. They also seem to target specific plants in relation with particular dishes (see § 3.1). The same tendencies have been observed in the Pacific Northwest area for instance (Poe et al. 2014) or in French Guyana (Tareau et al. 2019). Indeed, international migrations and population flows between rural and urban spaces have contributed to the circulation of knowledge and practices. This is the case for many Asian populations, as it has been shown on the circulation of plants and know-how by migrants between Laos and the United States (Bochaton 2019).

3. A low-key subversive practice for urban policies of nature management

3.1. *A variety of ecological motivations*

The motivations of the foragers vary accordingly with the profiles described in §2.2, thus offering a large panel of ecological mindsets. Figure 7.3 gives a synthetic view of the observed trends. They fall into two main categories: on the one hand, white, educated foragers who are integrated into metropolitan life; for them, foraging plants echoes rewilding practices (Dalgalarrondo and Fournier 2020). On the other hand, foragers of immigrant origin are less privileged and more marginalized from the metropolis but possess an inherited knowledge of plants and their uses. These two groups show very different motivations, and differences can also be observed within these two categories.

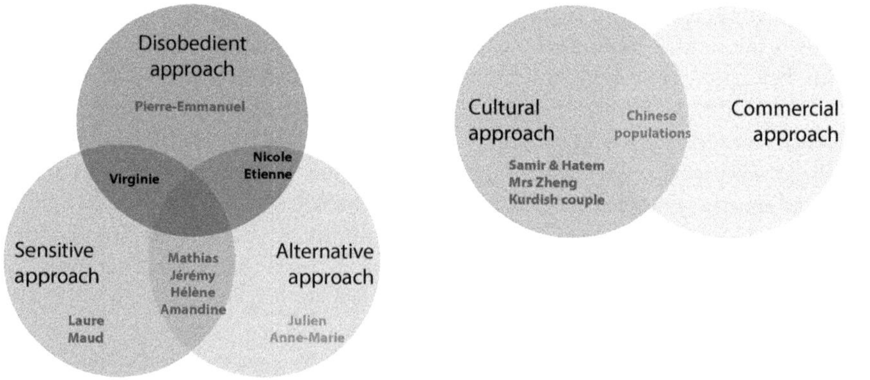

Figure 7.3 Categorization of foragers depending on their motivation
Authors: Roussel and Paddeu.

The first group of gatherers shares a common socio-cultural background and the same trajectory of rediscovery of plants in the city through a new personal awareness. The motivations behind this awareness vary. Three approaches can be identified, some of which overlap: (a) a sensitive approach that engages senses and the body, (b) an alternative approach that seeks to transform urban life and land uses and (c) a disobedient approach that transgresses land rights. The first approach echoes a rewilding of the existence and the body itself, whereas the second and the third reveal a more political form of rewilding.

a) In the first case, gatherers like Laure or Maud show a sensitive attachment to living things, echoing a spiritual quest, close to a form of personal development. Laure, 47, a painting restorer, holds a master's degree in physical methods applied to archaeology and museography, and practices yoga. She describes her connection with the living as "*an inner call*":

> *Afterwards, these are things you forget, you work, you live in the city, and then in 2016, there was a kind of big inner upheaval on many levels, big disillusionment too, really crash! And I had to find something to help me get back up again in a way, and afterwards it was really completely intuitive, there was nothing decided intellectually.*

She now collects for food and medicinal purposes, and also makes herbariums and botanical drawings, in an approach she describes as "*creative*." When she talks about the Black elder (*Sambucus nigra*), she says she is "*in love*" with it. She uses the flower to make syrup or dessert cream: "*It's floral, sweet, there's something very soft, very spring-like, very fresh*" she says. More generally, the relationship with living things is an opportunity for her to get to know "*my neighbors, they were friends*." Maud, 31, who has a master's degree in medieval history, decided one day to change her path by training as an herbalist. She has since founded an organization that organizes food and landscape events based on the theme of gathering. Her body is tattooed with willow branches because "*they embody resilience*," and she explains, echoing Laure's words: "*Plants, the more you get to know them, the more you feel affinity*." They are "*a way of reassuring yourself while facing anxiety*." The issues of health and care for the body, but also for the environment, are a recurrent criterion for this category of gatherers. In a more activist vein, Virginie mentions:

> *I wanted to discover medicinal plants, so really the popular medicine side, what use you can make of plants, as much to take care of yourself as to take care of animals.*

b) In the second case, foragers develop a reflection on how to live in the city. They question the sharing of resources, community life and the ecological enhancement of urban areas. Their involvement takes various forms from the more integrated to the urban economy and governance to more marginal forms. Julien is the founder of an organization that works with local authorities to implement urban greening projects. In this context, he participates in the "*beautification of neighborhoods*" while promoting edible plants within a "*nourishing ecosystem.*" He adds social objectives: "*the idea is that these spaces can benefit people who are in need, we need fresh food for everyone.*" More generally, it is a question of "*accompanying the inhabitants.*" It shows a concern for the sharing and preservation of resources, in line with the ideal of the commons:

> *which can be dangerous if you spread it massively, well I can understand that secrets are kept in the sense that if it is spread too much and not coordinated, it can lead to what is called "the tragedy of the commons." For it not to fall into tragedy, this resource must be managed by a group that guarantees that it is not squandered.*

Mathias integrates gathering to a mainstream economy that generates sufficient income to support a business. His rising Parisian start-up specialized in gathering plants organizes excursions for a day and offer supplement information on the web with tutorials and videos. Even if this type of stakeholders defends an ecological approach, it is a testimony of the merchandizing opportunities of the sector. It contrasts with others who keep the economical dimension at distance, such as Amandine, an auto-entrepreneur still registered at the French employment institution ("Pole Emplois") and benefit from a family apartment without charges, co-founder of a NGO that organizes excursions and wild plant meals, who states: "*We set the price at 35 euros per day, which is not that expensive. At first we thought it was 40 euros, but I'd rather have more people coming and it being cheaper than the other way around.*" But she still acknowledges: "*Now, yes, there are the realities of life that we may need money.*" Although she does not manage to make a living out of it, she belongs to a category of persons that share the secret ideal of doing so. These training are often expensive: 690 euros for the online training of the Chemin de la Nature or 700 euros for five days with the pioneer ethnobotanist François Couplan, adding food and transport, two examples of the opportunities of merchandizing of the sector.

c) Finally, we have identified a third category of populations whose practice is about challenging the capitalist system. The foragers do not hesitate to break out of the established framework of property and rules to defend another political vision. Pierre-Emmanuel organizes botanical events in parks through an organization he created. Describing himself as an "*activist,*" he states:

> *everything I do is part of a global thing in society, you see here I shelter young unaccompanied minors, I rent squats to do this, I also do things with wild plants in squats, to put on balconies.*

More generally:

> *In fact, my activity of popularising science, I do it so that people can become emancipated, I don't do it just for the money, besides I would stop if it was for that, because it's not profitable!*

While having relations with the municipalities, Etienne's organization shows its support for ecological resistance movements such as the one in Notre-Dame-des-Landes against a new airport project. In Montmagny, he brings people from poor neighborhoods to forage currants, raspberries and blackberries at the edge of community gardens where they tend a parcel. In this category, property trespassing is seen as a way to give access to resources to underprivileged people, or to valorize products that have been abandoned. For example, Nicole maintains abandoned plots of orchards that do not belong to her on the argument that this resource must be preserved in the context of soil artificialization and the ecological crisis: "*That's why, for me, first of all, it's a green heritage to be safeguarded.*" This questioning of the system can also be part of societal dynamics. For example, Virginie, already mentioned in the first category, sees her approach as part of a "*feminist action*" by taking an interest in "*plants to heal women.*"

Figure 7.3 shows a second set of populations of immigrant origin, more modest and with more discreet practices. These are part of a cultural heritage based on the use of plants, which has been transmitted or directly imported from the countries of origin. Thus, Samir and Hatem, two inhabitants of the municipality of Epinay, construction workers, collect thistles to make couscous in the Kabyle tradition. In La Courneuve, a couple of Kurdish origins picks the flowers of the Lime trees (*Tillia cordata*) to make infusions like those they used to drink in Turkey. Mrs. Zheng, already mentioned, working in the 18th *arrondissement* of Paris, uses the culinary knowledge she learned in China with plants from her Parisian neighborhood. For instance, she cooked a tasty dish of rice and Wild lettuce (*Lactuca virosa*) fried in soy sauce for us. For other dishes, she also uses Dandelions (*Taraxacum officinale*) and seeds of the Japanese Sophora (*Sophora japonica*), a tree planted by the city of Paris but originally from China. Many respondents of our survey mentioned the gathering habits of Chinese population. They are supposed not only to forage for private uses but also to provide resources to catering chains, and thus to ensure an income.[8]

3.2. *Standardized green space management policies*

Our results provide a joint response to the second and third hypotheses formulated in the introduction: the diversity of the ecological approaches of the gatherers does not fit in with the standardized approaches of the public authorities. Gatherers share an approach to plants based on a "relational ecologies of belonging" (Poe et al. 2013). This manifests in a sensitive connection with plants through identifying, touching and smelling. As a direct material interaction with plants allowing a sense of appropriation of place and the living world, foraging contradicts the dominant paradigm of biodiversity preservation and enhancement where the living is to be kept at distance from users and taken care of only by experts. For instance, in contradiction with experts, urban foraging contributes to the recognition that invasive and non-native species may provide ecological and

8 For the moment, we have not been able to explore this commercial dimension, which will be the subject of future surveys.

social benefits. Finally, whether they are aware of it or not, foragers act as low-key disruptors of the existing urban models and management regime of green spaces.

Although in most cases it is mainly tolerated, within the legal texts of various green spaces, foraging is an officially prohibited practice, sometimes leading to seizures and official reports, for instance, in the Parc Georges Valbon in La Courneuve. Foraging plants is disregarded by green spaces managers, such as urban gardeners and rangers who are responsible for conforming vegetal spaces to their attributed functions (such as recreational, aesthetic and ecological). And they have to deal with very limited educational support to integrate new practices. They perceive gathering plants as an ecological pressure on resources; as a source of disorder on landscape design; and as a destructive practice on public goods, sometimes assimilated to petty larceny. Generally, it challenges a sense of ownership, control and legal responsibility of green public spaces.

A central issue with respect to foraging hence remains access to land and to resource. While foraging was associated with common property regimes in the pre-modern world (Svizzero 2016), common property regimes are scarce in the capitalist world and threatened by privatization. Exclusive property rights have been generalized, especially in the Global North. This shift translates into fences, security guards and regular patrols, and into changes in landscaping design (Hurley et al. 2015). Urban picking requires permanent adaptability on the part of the gatherers, and the ability to overcome prohibitions and physical barriers. Foragers therefore rely on a higher reason that guides their actions: political, ecological, sensitive or linked to a traditional cultural anchorage. Some contexts, such as field courses and workshops, are an opportunity to legitimize the activity, whereas other populations are more exposed to stigma, particularly those of immigrant origin who are less integrated. For them, the suspicion of theft or degradation is stronger. But we do not know yet specifically how gatherers are differently treated depending on class, race or gender.

Rules on the books are prone to ambiguous interpretation or are rarely enforced, revealing tolerance toward these practices, and sometimes their legal recognition. For instance, if institutional rules often prohibit the collection of plants on public lands, some park managers recognize the benefits of environmental awareness and ecological knowledge that come from foraging. In some cases, authorities create convenience zones, institutionalizing informal practices (McLain et al. 2014). Some city governments are slowly beginning to support gathering practices, often through partnerships with civil society groups. There is a renewed interest in the role of food production within some park spaces, including cases of creating orchards and agroecological landscapes within city parks and planting food crops within street easements (Blanc and Paddeu 2018; McLain et al. 2012). It is a way of thinking about new modalities of urban planning that include foraging practices, whether through the notion of "edible infrastructure" (Russo et al. 2017), through associative and municipal practices such as in Seattle (McLain et al. 2012, 2014), or municipalities we investigated such Arcueil or Paris, which open up a reflection on park and nature governance and management (Krinsky and Simonet 2017). For instance, residents are often criticized by professional gardeners for not knowing how to pick, as our interviews with park managers show. When municipalities take up the issue of edible plants in the city, the framework they define can appear to be a forced commonality: for example, when the municipality of Arcueil invites passers-by to take hold of the fruits from public community gardens, this creates conflicts with the gardeners involved. The practice seems to be difficult to govern from the top.

The types of practices that are fostered reveal uneven treatments of varied forms of foraging. Some of these programs appear to maintain a two-tier system, in which only a limited number of purposefully planted species and deliberately maintained spaces are recognized, while many other species, areas, and practices are not. Tree fruit harvesting for charity organizations is supported by municipalities and advertised as "innovative" or "best" practices, while unauthorized foraging is stigmatized and discouraged, especially when performed by and with marginalized groups such as migrant population. Yet these arrangements are intrinsically transient and subject to uneven abilities to mobilize resources and capitals. In the background, it reveals different levels of appropriation of the ecological questions by foragers as well as new ways of investigating the question of the governance of the commons.

Conclusion

Looking at plant foraging in the city is a gate to discuss the possibilities of living with nature in an urban environment. These possibilities refer, on the one hand, to the availability of green spaces and plants in very constrained environments. Our study shows that gatherers target a wide variety of green spaces, some close by, some far away, some abandoned, some highly maintained, and take advantage of recent policies in favor of biodiversity in particular. The possibilities also relate to the ability of city dwellers to forage. We can observe a diversity of profiles as well as a diversity of motivations, including a wide range of ecological sensitivities. There is not one form of ecological relationship with nature in the city, but several. Finally, possibilities refer to the regulatory and land barriers to access these resources. By showing a relational approach to the living world, gatherers disrupt the more normative approach of the public authorities, which alternates aesthetic and nature conservation objectives. Harvesting thus highlights the powers at work between stakeholders in the field of green space management.

References

Blanc, Nathalie, and Flaminia Paddeu. 2018. "L'environnementalisme ordinaire. Transformer l'espace public métropolitain à bas bruit?" *EspacesTemps.net. Revue électronique des sciences humaines et sociales*. www.espacestemps.net/articles/lenvironnementalisme-ordinaire-transformer-lespace-public-metropolitain-a-bas-bruit/

Bochaton, Audrey. 2019. "Intertwined Therapeutic Mobilities: Knowledge, Plants, Healers on the Move between Laos and the U.S." *Mobilities* *14*(1): 54–70. https://doi.org/10.1080/17450101.2018.1522878

Bromberger, Christian, and Gérard Lenclud. 1982. "La chasse et la cueillette aujourd'hui. Un champ de recherche anthropologique ?" *Études Rurales* 87(1). www.persee.fr/issue/rural_0014-2182_1982_num_87_1

Clément, Gilles. 2006. *Le jardin en mouvement. De la vallée au jardin planétaire*. Sens et Tonka.

Dalgalarrondo, Sébastien, and Tristan Fournier. 2020. *L'Utopie Sauvage*. Paris: Les Arènes.

Flegeau, Morgane, and Fabien Roussel. 2021. "L'improbable ceinture verte: aménagements et dynamiques socio-territoriales des espaces végétalisés des pourtours de l'agglomération parisienne." In *Les nouveaux paradigmes du projet? Transition, adaptation, résilience*. Edited by Helga-Jane Scarwell and Philippe Deboudt, 215–237. Bruxelles, Belgium: Peter Lang.

Francis, Robert A., and Michael A. Chadwick. 2012. "What Makes a Species Synurbic?" *Applied Geography* 32(2): 514–521. https://doi.org/10.1016/j.apgeog.2011.06.013

Gandy, Matthew, and Sandra Jasper, eds. 2020. *The Botanical City*. Berlin: Jovis.

Gilli, Frédéric. 2014. *Grand Paris. L'émergence d'une métropole*. Paris: Presses de Sciences Po.

Grésillon, Étienne, Frédéric Alexandre, and Bertrand Sajaloli. 2016. *La France des marges*. Paris: Armand Colin.

Hurley, Patrick T., Maria R. Emery, Rebecca J. McLain, Melissa R. Poe, Brain Grabbatin, and Cari L. Goetcheus. 2015. "Whose Urban Forest? The Political Ecology of Foraging Urban Nontimber Forest Products." In *Sustainability in the Global City: Myth and Practice*. Edited by Cindy Isenhour, Gary McDonogh, and Melissa Checker, 187–212. New York: Cambridge University Press. www.nrs.fs.fed.us/pubs/48695

Kangas, Kari, and Pasi Markkanen. 2001. "Factors Affecting Participation in Wild Berry Picking by Rural and Urban Dwellers." *Silva Fennica* 35(4): 487–495. https://doi.org/10.14214/sf.582

Kaoma, Humphrey, and Charlie M. Shackleton. 2014. "Collection of Urban Tree Products by Households in Poorer Residential Areas of Three South African Towns." *Urban Forestry & Urban Greening* 13(2): 244–252. https://doi.org/10.1016/j.ufug.2014.02.002

Krinsky, John, and Maud Simonet. 2017. *Who Cleans the Park? Public Work and Urban Governance in New York City*. Chicago: The University of Chicago Press. https://press.uchicago.edu/ucp/books/book/chicago/W/bo25338739.html

Legrand, Marine. 2017. "Un décor comestible. Mise en ordre écologique des parcs urbains et collecte citadine de ressources alimentaires 'sauvages'." *Géographie et Cultures* 101: 97–117. https://doi.org/10.4000/gc.4975

Machon, Nathalie, ed. 2010. *Sauvages de ma rue*. Paris: Éditions Le Passage.

McLain, Rebecca J., Patrick T. Hurley, Marla R. Emery, and Melissa R. Poe. 2014. "Gathering 'Wild' Food in the City: Rethinking the Role of Foraging in Urban Ecosystem Planning and Management." *Local Environment* 19(2): 220–240. https://doi.org/10.1080/13549839.2013.841659

McLain, Rebecca J., Melissa R. Poe, Patrick T. Hurley, Joyce Lecompte-Mastenbrook, and Marla R. Emery. 2012. "Producing Edible Landscapes in Seattle's Urban Forest." *Urban Forestry & Urban Greening* 11(2): 187–194. https://doi.org/10.1016/j.ufug.2011.12.002

Millington, Nate. 2013. "Post-Industrial Imaginaries: Nature, Representation and Ruin in Detroit, Michigan." *International Journal of Urban and Regional Research* 37(1): 279–296. https://doi.org/10.1111/j.1468-2427.2012.01206.x

Muratet, Audrey, Myr Muratet, and Marie Pellaton. 2017. *Flore des friches urbaines*. Paris: Éditions Xavier Barral.

Paddeu, Flaminia. 2016. "D'un mouvement à l'autre: des luttes contestataires de justice environnementale aux pratiques alternatives de justice alimentaire ?" *Justice Spatiale/Spatial Justice* 9. www.jssj.org/article/dun-mouvement-a-lautre-des-luttes-contestataires-de-justice-environnementale-aux-pratiques-alternatives-de-justice-alimentaire/

Paddeu, Flaminia. 2019. "Waste, Weeds, and Wild Food." *EchoGéo* 47. http://journals.openedition.org/echogeo/16623. https://doi.org/10.4000/echogeo.16623

Paddeu, Flaminia. 2021. *Sous les pavés, la terre. Agricultures urbaines et résistances dans les métropoles*. Paris: Seuil.

Plieninger, Tobias, Claudia Bieling, Nora Fagerholm, Anja Byg, Tibor Hartel, Patrick Hurley, César A. López-Santiago, Nidhi Nagabhatla, Elise Oteros-Rozas, Christopher R. Raymond, Dan van der Horst, and Lynn Huntsinger. 2015. "The Role of Cultural Ecosystem Services in Landscape Management and Planning." *Current Opinion in Environmental Sustainability 14*: 28–33. https://doi.org/10.1016/j.cosust.2015.02.006

Poe, Melissa R., Rebecca J. McClain, Marla Emery, and Patrick Hurley. 2013. "Urban Forest Justice and the Rights to Wild Foods, Medicines, and Materials in the City." *Human Ecology 41*(3): 409–422. https://doi.org/10.1007/s10745-013-9572-1

Poe, Melissa R., Joyce LeCompte, Rebecca J. McLain, and Patrick Hurley. 2014. "Urban Foraging and the Relational Ecologies of Belonging." *Social & Cultural Geography 15*(8): 901–19. https://doi.org/10.1080/14649365.2014.908232.

Redzić, Sulejman. 2010. "Use of Wild and Semi-wild Edible Plants in Nutrition and Survival of People in 1430 Days of Siege of Sarajevo During the War in Bosnia and Herzegovina (1992–1995)." *Collegium Antropologicum 34*(2): 551–570.

Roussel, F and Frédéric Alexandre. 2021. "Landscape Ecological Enhancement and Environmental Inequalities in Peri-urban Areas, Using Flora as a Socio-ecological Indicator – The Case of the Greater Paris Area." *Landscape and Urban Planning 210*. https://doi.org/10.1016/j.landurbplan.2021.104062

Russo, Alessio, Francisco J. Escobedo, Giuseppe T. Cirella, and Stefan Zerbe. 2017. "Edible Green Infrastructure: An Approach and Review of Provisioning Ecosystem Services and Disservices in Urban Environments." *Agriculture, Ecosystems & Environment 242*: 53–66. https://doi.org/10.1016/j.agee.2017.03.026

Shackleton, Charlie M., Patrick T. Hurley, Annika C. Dahlberg, Maria R. Emery, and Harini Nagendra. 2017. "Urban Foraging: A Ubiquitous Human Practice Overlooked by Urban Planners, Policy, and Research." *Sustainability* 9(10): 1884. https://doi.org/10.3390/su9101884

Svizzero, Serge. 2016. "Foraging Wild Resources: Evolving Goals of an Ubiquitous Human Behavior." *Anthropology* 4(1): 1–9. https://doi.org/10.4172/2332-0915.1000161.

Synk, Colleen M., Brent F. Kim, Charles A. Davis, James Harding, Virginia Rogers, Patrick T. Hurley, Marla R. Emery, and Keeve E. Nachman. 2017. "Gathering Baltimore's Bounty: Characterizing Behaviors, Motivations, and Barriers of Foragers in an Urban Ecosystem." *Urban Forestry & Urban Greening 28*: 97–102. https://doi.org/10.1016/j.ufug.2017.10.007

Tareau, Marc-Alexandre, Lucie Dejouhanet, Guillaume Odonne, Marianne Palisse, and Clarisse Ansoe. 2019. "Penser la cueillette de plantes médicinales sauvages dans des sociétés en transition: le cas guyanais." *EchoGéo 47*. https://doi.org/10.4000/echogeo.16679

Terada, Toru, Makoto Yokohari, Jay Bolthouse, and Nobuhiko Tanaka. 2010. "Refueling Satoyama Woodland Restoration in Japan: Enhancing Restoration Practice and Experiences through Woodfuel Utilization." *Nature and Culture* 5(3): 251–276. https://doi.org/10.3167/nc.2010.050303

Vanuxem, Sarah. 2021. "Du droit de déambuler. Le paysage comme lieu de passage." *Les Cahiers de l'École de Blois 19*: 10–19.

von Hoffen, Laura Pauline, and Ina Säumel, I. 2014. "Orchards for Edible Cities: Cadmium and Lead Content in Nuts, Berries, Pome and Stone Fruits Harvested within the Inner City Neighbourhoods in Berlin, Germany." *Ecotoxicology and Environmental Safety 101*: 233–239. https://doi.org/10.1016/j.ecoenv.2013.11.023

Weeks, Claire, Michelle Croasdale, Mark Osborne, Leisha Hewitt, Patrick Miller, Paul Robb, Malcom Baxter, Paul Warriss, and Toby Knowles. 2006. "Multi-element Survey of Wild Edible Fungi and Blackberries in the UK." *Food Additives & Contaminants* 23(2): 140–147. https://doi.org/10.1080/02652030500386184

8 Urban scavenging

Another way to rewild the self in the city[1]

Sabine Mégarbané

In a 2019 article, geographer Flaminia Paddeu, who is interested in the many gleaning-type practices in urban zones in developed countries, found an approach based on a "'forager-scavenger continuum' instead of opposing foraging for wild food to scavenging for discarded food" (Paddeu 2019). After a brief history of gathering and salvaging practices, followed by a presentation of their contemporary forms, this commentary will take a closer look at the similarities between foraging for plants and informal scavenging for unsold[2] food. Taking up the framework of the chapter by Fabien Roussel and Flaminia Paddeu, I wish to show how the transformation of urban gathering practices constitutes an original form of self-rewilding.

A "forager-scavenger continuum" rooted in a common history

Historically, practices of gathering, foraging and scavenging, which Marcel Mauss (1926) described as "industries of acquisition," provide food resources and economic resources for the poorest populations. In addition to gathering and foraging activities, which could be practiced relatively freely for many centuries, there is also a right of use intended to provide the poorest groups with a source of food: the right to glean. Mentioned early on in French law, the right to glean declined from the seventeenth century and particularly in the nineteenth century, as the rise of private property, the industrialization of France, and the growing urbanization of its population coincided. In cities, although gleaning ceased, other practices for gathering discarded or leftover items appeared, for example, ragpicking. The daily work of ragpickers was to collect and sort people's garbage, keeping the urban metabolism running and maintaining cooperation and cohesion between city, countryside and industry in exchange for payment (Barles 2005). Recent years have seen a gleaning-type practice reemerge in urban areas. In cities, after markets finish, people are often seen salvaging food. They "humbly stoop" (Varda 2000) in

1 Translated by Cadenza Academic Translations.

2 In this commentary, I will use the terms "unsold" and "waste" interchangeably, even though they are not equivalent and do not refer to the same material situation: the first can be reinserted into the commercial cycle, while the second is excluded from it.

DOI: 10.4324/9781003243496-16

a motion similar to the ancestral figure of the gatherer. The practice is also being renewed and transformed by the arrival of new participants, who are motivated less by economic necessity than by their political ideals.

The trajectory of the practice of foraging is similar to that of gleaning. Once a widespread practice among the lower classes and a common way of obtaining plant resources, foraging decreased from the Middle Ages, particularly because of the increasing privatization of spaces. Over the centuries, the practice declined and became restricted to certain geographical areas and social strata, gradually turning into an "activity of the poor" for those whose survival still depended on it (Coujard 1982).[3] Today, as shown by Fabien Roussel and Flaminia Paddeu, the practice of foraging is reemerging in the interstices of the city and in the natural and green spaces that appear despite the density of the urban environment. Urban foraging is therefore used by people seeking extra food to supplement their everyday consumption or additional income, and by those for whom it helps to express a certain closeness with nature, reconnect with an element of wild living, take ownership of their environment or simply enjoy themselves.

Unrestricted scavenging?

Salvaging unsold food, just like urban foraging, is attracting more and more people who are drawn to this practice for reasons other than pure economic necessity. At the end of the 1990s and in the early twenty-first century, the term *freegan* appeared, followed by the movement of the same name, in anarchist and punk circles in the United States. This movement encourages autonomization, based on the observation that there can be no such thing as ethical consumption in the capitalist system and that any participation in this system is inevitably detrimental to humans, animals and the planet. Also laying down the foundations of the movement, the manifesto *Why Freegan? An Attack on Consumption – In Defense of Donuts* proposed the adoption of several practices that would allow people to participate as little as possible in the capitalist system. It suggested various measures, including squatting for accommodation, hitchhiking or cycling for travel, voluntary unemployment, and even theft and dumpster diving to meet one's needs. The latter, dumpster diving to salvage food or other goods, is the practice most strongly emphasized by the freegan community. Out of the range of practices used, it is also generally the first that people adopt. This is paradoxical because dumpster diving is, a priori, an unattractive activity. However, it is the practice that gets the most attention from freegans and the media (Barnard 2011).

Although dumpster diving and salvaging waste are still widely stigmatized, it now seems that these practices are no longer restricted to the poor and socially excluded. In fact, in their research on urban food gathering, Olivier, Nicolaï, and Riffaut (2010) show the presence of new groups competing in the scavenging

3 Translator's note: Unless otherwise stated, all translations of cited foreign-language material in this commentary are our own.

spaces explored. These populations have higher levels of cultural capital and social integration than the populations who traditionally engage in such practices. These new urban scavengers primarily highlight ecological and political motives for salvaging unsold food. They form a relatively unstructured movement, at least in France, but with a certain presence on social networks. Online, several groups dedicated to freeganism bring together experienced and aspiring urban scavengers. The online exchanges are an opportunity to share advice on salvaging, request help, post photographs of particularly impressive finds, and give or barter food. It was via these groups that I was able to find and meet my survey participants. Using the internet and choosing people who at least identified with the term "freegan" was a deliberate choice to explore the issue of salvaging unsold food with people who were a priori at ease with the practice, and for whom the survey would not feel stigmatizing. Far from allowing themselves to be stigmatized, the people I met would "reverse" or "redirect" the stigma (Nguyen, Chen, and Mukherjee 2014). First, they emphasized that they had chosen to take up this practice completely freely and willingly, and second, they adopted a hierarchy of values that reversed that of dominant groups, redirecting the shame toward them. The survey participants' ability to counter the stigmatization connected with dumpster diving and proximity to waste should be interpreted in the light of their socioeconomic status. Most of the people I met were under 30 years old and not in salaried employment at the time of the survey. Half of them were living on state aid or had incomes below the poverty line, even though all of them were from middle-class backgrounds and had received or were in higher education. They were living in or on the outskirts of French cities. They did not deny their economic motivation when interviewed, but they always presented it as secondary, coming after political or environmental motives, or their enjoyment of the practice. Freegans are therefore seemingly characterized by their (relatively) high level of cultural capital and low level of economic capital. According to Mathieu Grossetête (2019), this category, with its environmental commitment, displays a certain "interest in disinterest," experiencing material deprivation as voluntary rather than as a forced sacrifice. This succinct presentation of the survey participants and their similarities with some of the urban foragers met by Flaminia Paddeu and Fabien Roussel lays the foundations for me to explore the multiple connections between foraging and scavenging, and to examine how the latter, against all expectations, is a manifestation of a connection to nature and an expression of self-rewilding.

Rewilding the self in the city

In a 1991 article, Bernard Picon defines foraging activities above all by the pleasure they give to individuals, whatever their motivations. According to Picon, gathering food, like hunting and fishing, is a chance for those involved to "play the role of a '*homo naturalis*' seeking hard-to-access natural resources" (Picon 1991, 88). It therefore becomes the opportunity for a heroization of the self as a "man of the woods" capable of searching for and finding food and useful items thanks to his knowledge and skills.

Knowing and belonging within a territory

Just as foraging requires a detailed and intimate knowledge of the area and the products it has to offer, scavenging for food requires skill in searching and an understanding of the workings of the city. Consequently, just like sites for picking mushrooms (a delicate, tasty and often expensive food) are hard to find (Bouverot-Rothacker 1982), scavenging sites can also be difficult to unearth. On social networks online, it is regularly requested that people not give the exact address of the places where they have found products, for fear that the site might become a victim of its success, but also because many feel that searching for a good salvaging site is an integral part of the practice. Searching is also a chance to get to know and take ownership of an area, a part of the city, or a neighborhood, becoming familiar with its stores, its garbage collection times, the best waste receptacles, and their accessibility.

> You search, you find nothing, you give up, you search, you find nothing, I remember spending whole afternoons outside an organic store with a book to see how they put out their garbage . . . in fact, all you need is for things to fall into place once, one successful salvage, to continue, and then to start to understand how a city works too. . . . Because when you start from scratch, you need to understand how it works, find out how many times a week a certain store puts its trash out, whether they have a compactor or not: there are lots of questions. You need to be a bit organized, and I think that, yeah, it took six months to get there. . . . We realized it wasn't the middle-class market in the city center that discarded the most, it was the working-class market in the middle of a deprived area that threw lots away. Because it was full of traders who bought lots of products in bulk, who didn't care what they bought, or it was by business, so they kind of did what they could.
>
> – Emmanuel, 28, studied as an engineer (five years at university), freegan for seven years, unemployed

Like foragers whose practices sketch out what Fabien Roussel and Flaminia Paddeu call a "pattern of gathering," freegan scavenging implicitly represents a certain view of the city in which the notions of liberty, autonomy and food justice are central. Similar to Lefebvre's concept of the "right to the city"[4] (1968), scavenging for unsold food transforms and shapes the city in line with the freegan ideals of anticapitalism, green living and solidarity. In this, urban scavenging sites may represent what David Harvey (2011), using a concept from Foucault, calls a "heterotopia": a place where "something different" is possible. They are spaces where in addition to the existence of working-class and subsistence practices, other ways of being in the city and living on the margins of the capitalist system are invented.

4 Henri Lefebvre defines the right to the city not simply as a right to housing, but as a right to inhabit the city, to take ownership of it and to transform it. He advocates a true coproduction of the urban space by its inhabitants, through their everyday practices, by twisting and modifying spaces according to their needs and uses.

Just like foraging, scavenging is a chance to walk, wander and stroll. Urban scavengers are guided by opportunity. Sometimes they might prefer to scavenge locally, or they might prefer to use a vehicle to travel to more distant sites, such as commercial zones on the outskirts of cities, in the hope of finding larger quantities and more variety.

Reconnecting with our wild instincts

It is easy to imagine how foraging can be demanding for the body, which not only is tested but also becomes a tool in service of the practice, able to distinguish the edible from the toxic. This physicality is also seen in scavenging. The body is used to handle garbage cans and dumpsters, moving them, tipping them up, and half (or even fully) diving into them to search inside. Furthermore, dumpster diving offers a much less sterile and ordered context for obtaining supplies than food stores. The body, along with its senses, becomes the main mediator when searching through garbage.

> An adventurous little moment, so feeling slightly proud, you get your hands in there, it's the manual aspect too, it's not back to the earth, it's back to the dumpster, but you also feel like you've made an effort and got something out of it.
>
> – Carole, 23, studied as an engineer (five years at university), beginner scavenger, performing voluntary civilian service

Touch, sight, smell, taste and hearing: all the senses are used in dumpster diving. In supermarket aisles, most consumers cursorily check the expiry date before putting a product in their basket, but dumpster diving, just like foraging, calls for a little more care. Salvaging unsold food requires specific measures to be taken. Dumpster diving is very different to shopping. Divers rely on their senses to identify any "faults" with food that might make it unfit for human consumption. The sensory organs are particularly important for deciding whether food is still edible or not. In a way, and precisely because those scavenging for unsold food cannot just choose a product from the shelf and instead need to examine it thoroughly, freegans are closer to the materiality of their food. Like for hunters and gatherers, obtaining this food takes time (lots of it), work, effort and commitment. Where the forager can generally rely on their knowledge and their ability to identify and tell the difference between the species they pick, the urban scavenger must make sure that the food they find is edible, and that it has not become unfit for consumption, been contaminated by other products in the dumpster, or been eaten by various rodents. It is often the senses of smell and taste that allow them to discriminate and decide precisely whether a product is edible once it has passed the visual test.

> Last year, the supermarket had thrown shrimp in the garbage. Shrimp are fine as long as they don't smell! Except there was one that smelled in there, and I know it, it's stupid, seafood is easy: When the head starts to smell, you can eat the head that day, and the next day you can't eat the head, the body starts

> to smell, you can eat body that day, the day after the head starts to smell, and two days after, you should stop. There was a whole panful there, and I could tell the head had a bit of a smell, but I ate the things anyway, so I had diarrhea in the night, and the next morning it had passed, I just spent the night on the can and that was all. Meat and fish, when it's really bad, it smells, you just DON'T want to touch it.
>
> – Maya, 25, doctoral student studying science (eight years at university), freegan for four years

Urban scavengers therefore need to check that the food they find is safe to eat. They do this using their senses, particularly smell and taste. However, ensuring that the products are fit for consumption is not the only step required before eating them: to be upgraded from waste (back) to edible, the food must go through several stages of sorting and processing, particularly by cooking.

Redefining edibility

In their chapter, Fabien Roussel and Flaminia Paddeu show that foraging for wild plants in towns and cities is redefining the usual boundaries between good plants and bad plants. Ultimately, they assert that this distinction, which depends heavily on the potential utility of the plant, tells us more about the people who pick it than about the plant being harvested. This distinction is in fact largely influenced by socioeconomic and cultural differences, and by the individual's knowledge about the plant's characteristics, and therefore about its usage value or economic value. Similarly, waste salvaging demonstrates the "fluidity" (de Laet and Mol 2000) of the status of waste and its transitory nature. Waste can be freely reappropriated under French law, and various processing operations can help to restore value to it. The products salvaged from garbage are rarely consumed as they are found. Generally, they go through various stages before they are truly considered food again. Beyond the fact that salvaging unsold food requires a particular timetable with unusual hours that are often not especially compatible with salaried work, it also imposes a certain rhythm. Because the products salvaged are highly perishable and already quite old or ripe, it is often necessary to immediately sort through the food, clean it and do some basic preparation so that it will last longer before it goes bad.

> It took us a long time to search through the garbage, bring the food back, and "prepare" everything. I pre-chopped all the vegetables so I could freeze them. You have to sort the food too. When you find a basket of raspberries, half of the fruit is rotten, so you need to be meticulous. It's the same for salad. You remove the leaves that look bad, you keep the salad, so it's quite time-consuming.
>
> – Carole, 23, studied as an engineer (five years at university), beginner scavenger, performing voluntary civilian service

The stages of sorting, preparing, preserving and cooking are ways of at least partly erasing the food's previous garbage status. Cooking is the final step in converting the salvaged products into food that is both psychologically acceptable and good to eat. This means employing a familiar and identifiable "culinary repertoire" (Fischler 1990).

Growing stronger through exposure: immunoevolution as a solution

Consuming products found in waste receptacles, like eating foraged plants, presents a health risk. Certain measures and precautions are necessary to guard against this. However, this risk, this boundary between the edible and the toxic, can become a chance for the person eating the food to stretch their own capacities. Some want to expose their urban body, which they believe is weakened by an overly sterile environment, to various bacterial stimulations that might strengthen it thanks to the process of "immunoevolution." This term, used on the "Freegan Station" website (a reference website for the French freegan movement), is inspired by the concept of immunoevolution, which is used in particular in animal biology. Here, self-rewilding takes an animal-inspired turn. Immunoevolution refers to the adaptation of the immune system to prevent fatal attacks by microbial pathogens (Cooper 1992). The author of the French website adapts the term, defining immunoevolution as the immune system's capacity to grow stronger and more resistant, particularly through repeated exposure to bacteria by consuming expired products. The site's author emphasizes two key elements: The first is a desire to move away from modern food, which is considered sanitized and artificial. The second, the notion of immunoevolution as it is presented on the website, denotes a desire for "self-optimization" (Dalgalarrondo and Fournier 2019), which could also be described as "transhumanism with bare hands," to quote the name given to it by Dalgalarrondo and Fournier in their chapter on fasting (see Chapter 5). "Transhumanism" because of the promise to optimize our physiology by stimulating the immune system, and "with bare hands" because this self-optimization is achieved by activating a physiological potential via "technologies" that are accessible, because they are free and available. As in the case of fasting, the aim is "to do more with less."

> Still, I took it quite far, I liked pushing the limits of what was edible. So, generally, you judge by sight and smell, but yeah, I set myself some limits. . . . I think that might also be the reason I don't really get sick. . . . At one time, I often told myself "ah, if I eat things that have gone bad, it will strengthen my immune system," which isn't necessarily true and I have no idea because I'm not a doctor.
>
> – Emmanuel, 28, studied as an engineer (five years at university), freegan for seven years, unemployed

The concept of immunoevolution, although mentioned differently in other works (Barnard 2016), remains relatively anecdotal. However, it constitutes an additional argument in the assemblage of the promises of freeganism, meaning that the food's garbage status is no longer a fault to be hidden, but an opportunity for self-optimization thanks to the various molds that might have developed on it.

This commentary follows on from the work on foraging by Flaminia Paddeu and Fabien Roussel, by showing how scavenging for unsold food shares many characteristics with this first practice, and how, even though it is, a priori, an urban practice that is a far cry from the image we have of nature, it shows a connection with nature by reproducing actions, knowledge and ways of being in the world. Scavenging as practiced by freegans turns the body into both a tool in service of the practice and a support for their ideals. It allows them to feed themselves by salvaging, but it also helps them express their everyday environmental and anticapitalist commitment through the repetition of the practice and through the food they eat. Although unsold food is currently salvaged by various groups for whom urban scavenging is, to different extents, a means of subsistence, the practice is also subject to competition that might eventually cause its disappearance. It is jeopardized by movements to privatize waste (Guien 2019) and by the ever-more structured economy of food donations, which is being joined by a rapidly rising number of associations, businesses and state institutions in the bid to fight against food waste (Mourad 2018).

References

Barles, Sabine. 2005. *L'invention des déchets urbains. France 1790–1970*. Seyssel: Champs Vallon.

Barnard, Alex V. 2011. "'Waving the Banana' at Capitalism: Political Theater and Social Movement Strategy Among New York's 'Freegan' Dumpster Divers." *Ethnography 12*(4): 419–444.

Barnard, Alex V. 2016. "Making the City 'Second Nature': Freegan 'Dumpster Divers' and the Materiality of Morality." *American Journal of Sociology 121*(4): 1017–1050.

Bouverot-Rothacker, Anita. 1982. "Consommer l'espace sauvage." *Études Rurales* 87–88: 131–137.

Cooper, Edwin L. 1992. "Overview of Immunoevolution." *Italian Journal of Zoology* 59(2): 119–128.

Coujard, Jean-Louis. 1982. "La cueillette, pratique économique et pratique symbolique." *Études Rurales* 87–88: 261–266.

Dalgalarrondo, Sébastien, and Tristan Fournier. 2019. "Optimization as Morality, or the Routes of the Self." Translated by Rachel Gomme. *Ethnologie Française 176*(4): 639–651.

De Laet, Marianne, and Annemarie Mol. 2000. "The Zimbabwe Bush Pump: Mechanics of a Fluid Technology." *Social Studies of Sciences 30*(2): 225–263.

Fischler, Claude. 1990. *L'Homnivore*. Paris: Odile Jacob.

Grossetête, Matthieu. 2019. "Quand la distinction se met au vert. Conversion écologique des modes de vie et démarcations sociales." *Revue Française de Socio-Économie 22*(1): 85–105.

Guien, Jeanne. 2019. "À prendre ou à laisser. Ressources, gestes et corps de la récupératrice alimentaire en milieu urbain." *EchoGéo* 47, online. Accessed on January 30, 2022. http://journals.openedition.org.ezproxy.campus-condorcet.fr/echogeo/16521

Harvey, David. 2011. *Le capitalisme contre le droit à la ville: Néolibéralisme, urbanisation, résistances*. Paris: Éditions Amsterdam.
Lefebvre, Henri. 1968. *Le droit à la ville*. Paris: Anthropos.
Mauss, Marcel. 1926. *Manuel d'ethnographie*. http://classiques.uqac.ca/classiques/mauss_marcel/manuel_ethnographie/manuel_ethnographie.html
Mourad, Marie. 2018. *La lutte contre le gaspillage alimentaire en France et aux États-Unis. Mise en cause, mise en politique et mise en marché des excédents alimentaires*. PhD thesis, Sciences Po, Paris.
Nguyen, Hieu P., Steven Chen, and Sayantani Mukherjee. 2014. "Reverse Stigma in the Freegan Community." *Journal of Business Research* 67(9): 1877–1884.
Olivier, Chris, Chantal Nicolaï, and Hadrien Riffaut. 2010. *Glaneurs dans les villes. Étude monographique*. Paris: CerPhi.
Paddeu, Flaminia. 2019. "Waste, Weeds, and Wild Food." *EchoGéo* 47, online. Accessed September 6, 2021. https://journals.openedition.org/echogeo/16623#quotation.
Picon, Bernard. 1991. "Chasse, pêche, cueillette: Un même objet support d'attitudes et de pratiques sociales différenciées." *Sociétés Contemporaines* 8: 87–100.
Varda, Agnès. 2000. *The Gleaners and I*. Documentary film. 82 mins. Ciné-Tamaris.

Conversation 5. Between resistance and cracks: what the process of rewilding does to the gender order

9 The wild side of man

How animal metaphors shape masculine food practices and midlife transitions

Nicoletta Diasio and Vulca Fidolini

Introduction

This contribution aims to explore the trope of wildness in masculine age transitions that take place in midlife. More precisely, we will study how the animal metaphors in men's accounts of their own food practices participate in producing a particular relationship with body, age and gender. We will specifically focus on masculinities that are facing the effects of time, by looking at the power of the animal metaphors which our interviewees used to describe their food practices and their role in managing the ageing process. Reformulating the relationship to "nature" and "the wild" is a widely represented dimension in the ethnological literature on age transitions, and we want to examine how this theme resonates when matched with more recent work on masculinity and ageing, body and food habits. This paper also aims to contribute to the flourishing contemporary literature on how men's eating habits shape constructions of masculinity, ageing processes and bodily changes (Sumpter 2015; Mycek 2018; Lax and Mertig 2020).

Men and masculine food choices have been studied in the Anglo-Saxon literature, particularly to question the binary opposition between food that is considered "more masculine" and food perceived to be "more feminine" (Adams 1990; Roos, Prättälä, and Koski 2001; Sobal 2005). Eating meat, and especially eating red meat, seems to be a major factor in the identification processes by which men adopt a hegemonic and virile masculinity (Connell 1995), and is also linked to a showier, more aggressive and less thoughtful approach to food (Nath 2011). On the other hand, eating vegetables and fruit seems to be associated with more "feminine" food practices (Gough 2007; Julier and Lindenfeld 2005). However, it would seem that this slightly static dichotomy between "masculine" and "feminine" food practices is not a faithful representation of the social and gendered attitudes toward food (Fidolini 2021). This chapter aims to go beyond these fixed oppositions and to question the way the masculine models are variably created. We consider food habits as "technologies of the self" (Foucault 1988), as

> procedures . . . that are proposed or prescribed for individuals to fix, maintain or transform their identity, depending on a number of goals, and this thanks

DOI: 10.4324/9781003243496-18

> to the management by the self of the self, or through knowledge by the self of the self.
>
> (Foucault 1989, 134)

These technologies are not fixed once and for all, and they cannot be separated from gender and age identification processes.

Ageing can cause changes in the way men accomplish what we might call a "body project" (Schilling 1993). By using this concept, we refer to the body as an object that is continually perfected, governed and expressed through social practices, including food, from both individual and collective perspectives. While ageing and the relationship with the body has often been analyzed from the point of view of women (Diasio and Vinel 2007; Charlap 2019), men's bodies have been much less studied. In this contribution, looking at the way food practices and the relation to age and gender intersect will allow us to understand how men shape their body and their perception of it through their choices in eating habits. By adopting a particular food regime, physical activity or fasting, the interviewees both fight with the weaknesses and uncertainties of an ageing body, and shape their masculinity.

Questioning the relationship between age and gender through food practices has allowed us to focus on some ambivalent processes which can be observed nowadays in our societies. On the one hand, becoming older does not necessary mean acquiring power: think for instance of the many injunctions concerning a perfectly healthy body, being efficient, or being in good physical shape. These injunctions, which all refer to hegemonic models of masculinity (Connell 1995), are embodied much more strongly by young people. On the other hand, as Jeff Hearn has shown (Hearn 1995), the background against which mature age has been staged in public life often contains images of men who have accomplished things, successful men, or men who surpass their limits, for example, through physical or sports performances. This is a masculinity model that is shaped through a specific relationship to the self, which combines flexibility and control, self-discipline and personal development, the injunction to be oneself and at the same time to adhere to powerful and incorporated masculine models and norms (Connell 1995).

The question is therefore: how does ageing combine with food practices and expressions of masculinity? And how are the metaphors that are inspired by the animal world and particularly the world of wild animals applied to express age transitions, where gender identifications are also reformulated? Indeed, during our qualitative study, the men we interviewed clearly connected midlife and their food practices with animal metaphors, especially to renegotiate masculine gender norms. Such references to animality attracted our attention, because metaphors inspired by cycles in nature, especially those which refer to a return to the wild as a liminal moment of passage in biographical and social terms, are a central theme in anthropology. These symbolic resources give rise to a wide range of practices which, depending on the social and historic context, allow us to signify and establish age, gender and social position. Even though they may seem far removed from the urban practices of our interviewees, the vitality of these discourses incited us to question how pervasive they are and how this relates to the interviewees' characteristics. We therefore decided to focus on a masculine rhetoric that creates correspondences with animality and wildness to analyse how men experience midlife. How do their food practices – and specifically meat-based or vegetarian diets – both challenge and confirm traditional assumptions about how men eat, while at the same time re-shaping hegemonic and subordinate profiles of masculinity?

The pertinence of taking food into account when reflecting upon gender is confirmed by the central place food has in the European context. Indeed, the occurrence of several food scandals, the increased awareness of health risks related to food, the pervasiveness of chronic illnesses where food is a main factor, the evolution of food-related social practices, environmental concerns and an increased demand for different food production and consumption methods make food an area where public policies are particularly prevalent. In France, for instance, the National Programme for Nutrition and Health has since 2001 put out recommendations and public health campaigns that target groups considered to be "at risk" because of age, income, work or health status. Similar targets have been set in Italy, with the program *Linee guida per una sana alimentazione italiana* issued by the Ministry of Agriculture. In this way, food represents a space where norms and new moralities are produced (Depecker, Lhuissier, and Maurice 2013; Dalgalarrondo and Fournier 2019). As it brings together the regulation of the "individual body" and the control over the "body as a species" ("*corps-espèce*" Foucault 1997, 216), food is therefore at the core of contemporary subjectivation processes (Diasio 2004). And because of its political dimension, food today constitutes a battlefield, an area of confrontation, debate and social creativity. It should therefore not surprise us to find food at the center of a redefinition of masculinity and ages in life.

After presenting the empirical research on which this text, our methodology and our aims are based, this contribution will first focus on the importance of taking into account a number of critical aspects of adult age. Then, we will discuss the ways in which age transitions, midlife and a relation to animality can be thought together. This will allow us to explore the ways in which men's relationship to time and food allows animal metaphors to shape different profiles of adult masculinity. Such gender identifications will be examined to see how they intertwine with ageing processes, and how they lead men in becoming more involved in a daily practice which they consider central: food consumption. Indeed, the study on which this chapter is based deals with men's narratives on food practices in two European countries, France and Italy, where food habits occupy a central place in the national culture (Fischler and Masson 2008). Local traditions of food preparation are highly valued, and meat-based dishes are especially prized. Thus, considering the importance of commensality in these countries, we put forward the hypothesis that in France and Italy, the fact of adopting a diet that excludes specific products (like meat) or food preparation techniques (e.g., refusing to eat barbecued meat) might be more difficult, or might need to be justified, because of the norms that govern eating practices and their social frameworks.

Methodology

The findings of this chapter are based on qualitative research carried out between February 2017 and July 2018. The research was based on semi-directed recorded interviews with 29 men between the ages of 40 and 63 who live in the region of Grand-Est in France and the region of Tuscany in Italy. All the interviews were transcribed and then analyzed in order to identify qualitative patterns that were central or peripheral to the collected narratives. Interviews were carried out in public places, cafés, restaurants, places of work or in private spaces like private homes, depending on the preferences of the interviewees. Their names are all fictionalized. The average duration of the interviews was one-and-half hours.

The men who participated in the research had diverse social backgrounds, professions and levels of education (see the *Appendix: Table of Interviewee Characteristics* for more details). All of them also declared being heterosexual. Most of the interviewees eat meat, other are vegetarian or vegan, and some could be defined as "flexitarians": they are principally vegetarian, but, in some situations, particularly on occasions of commensality, they eat meat, fish and products derived from animals. The cases, which will be presented in this contribution, are the most emblematic and the most representative of the main outcomes of the study. The aim of the chapter is not to compare eating practices in France and Italy. The sample is not large enough to do so, and this was not the purpose of the interviews. Moreover, the number of meat-eaters and non-meat-eaters included in the study is not statistically representative of the percentages of these categories in the samples of the two regions. The choice of doing fieldwork in these two regions was principally based on the fact that we had already conducted research into food practices and gender constructions in Italy and France and specifically in these two regions, where meat consumption is traditionally very high, and is often used to identify with a virile masculinity (Méchin 1985; Meloni 2011). We met the interviewees either through interpersonal networks or via organizations whose members share similar food diets, such as vegetarians or vegans, or who endorse a particular model of consumption. Examples of the models and concepts which such organizations embrace include short food chains, direct buying from producers, slow food and the promotion of vegan or vegetarian restaurants. One obstacle worth mentioning which we faced in carrying out the fieldwork are the difficulties we experienced in interviewing the men and women in the entourage of the interviewees, like wives, family members, work colleagues or friends. The fact of discussing the food practices of the interviewees with their peers or their partners undermined the relationship of trust established between the researchers and the interviewees. Considering that we risked jeopardizing the realization of the study, we chose to limit the interviews to men.

Sailing into a critical age

Just like masculinity (Bertrand et al. 2015), adult age has long been an unexplored and unformulated horizon in the social sciences (Blatterer 2007), a silent and obvious phase of life, a "normality" that is defined through comparisons with its "exceptions" (childhood, youth, old age). People between 40 and 60 represent an age category that is not very visible in social studies, mainly because it is difficult to define the exact frontiers of this stage in life: on the one hand, it is seen as a last prolongation of youth, and on the other hand, it is a preparation for old age.

"Mature age," "midlife" (Gullette 1998), "male menopause" (Hepworth and Featherstone 1998; Vainionpää 2005), sometimes also called the climacteric (Delanoë 2006) or the andropause (Vinel 2007): there are many different ways to critically define "an important transitional period, when attitudes, behaviors and practices which mark our social identity in our youth and early adulthood are challenged in the Western consumption culture" leading to the destiny of old age (Schwaiger 2009, 274–275). Studies have shown that what characterizes this period in life are the vague and flexible limits by which it is delineated – ranging between the age brackets of 30 and 70 (Vinel 2007, 205) – but also a medicalization and hormonalization of the masculine body, through a focus on its sexuality. The "vocabulary of the midlife crisis" (Featherstone and Hepworth 1985) indeed reveals a society that is permeated by the injunction to be young, to perform sexual and physical achievements, to care for one's appearance, and to have a body project that aims for health and fitness, in spite of ageing. All these features are also essential prerequisites for the actor's participation in contemporary consumer society (Parasecoli 2005).[1]

However, none of the terms which have been proposed by the scientific literature – mature age, midlife, andropause and the climacteric – seem to contain the complexity which characterizes this life stage in the narratives of our interviewees, or its ambivalence in relation to the way they experience their body's changes. By using the term "critical age," we here want to distance ourselves from definitions borrowed from the language of biology, as well as from any constructivist studies that identify the "50-year crisis" as a response to the injunctions of consumer society (Vinel 2007). The expression "critical age" is partly linked to an older question concerning the end of the period called "adult age": a question which goes back to the eighteenth century. The word "critical" is rooted in the Greek *kritikos*, or *krinein*, meaning "able to judge, to decide." In medical language it is often associated with the Latin word *dies*, day, in the sense of a "decisive, critical" stage which will determine the rest of the events, or to "qualify something which is marked by crisis."[2] The main aim of our lexical and conceptual choice is here to go back to the experience that our interviewees describe in their narratives when they refer to the uncertainty and vulnerability they feel in facing the adult phase of life.

1 It is interesting to observe that the importance conferred to sexuality is also confirmed by studies on ageing (Bessin and Blidon 2011). These studies look at the transition from adult age to the early stages of old age and especially investigate how sexual practices are renegotiated.

2 In French, according to the *Dictionnaire historique de la langue française Robert*, the word "crise," which is equivalent to the English word "crisis," only became more widespread in the eighteenth century (1762) in its current meaning of "that which decides over the fate of someone or something, which brings change."

Such feelings cause the development of specific body practices which are especially focused on monitoring food practices. As for other intense moments in an individual's existence, "the end of the first part of the adult life" (Lalive D'Epinai and Cavalli 2007, 51) constitutes a moment of truth where social norms on ageing manifest themselves clearly.

The word "critical" also refers to the thresholds that separate the different ages of life. Anthropologists have clearly demonstrated how mutations in an individual's position within the social space, and periods of crisis or social transformation, entail a risk of disorder and call into question the usual categories that confer meaning and order to life in society. Marginal situations free up potentialities that require new efforts of classification and ordering (Douglas 1970). Such moments of re-arrangement can give rise to profound fears of returning to the wild, among which a return to the animal state is one of the main symbols. The discourses and practices that draw on "nature," where nature is obviously always a construct and is signified through human action, thus constitute resources that express our entry into a new time in history and our passing of thresholds necessary to produce forms of humanity (Remotti 2002). Associating crisis with a return to the wild also reminds us how these procedures of categorization, ordering and "domestication" of reality are, like the process of taming an animal as described by Digard (1993), unstable, reversible and always under threat of disorder and informality. With the example of male puberal development, Marika Moisseeff thus shows, for instance, that

> the contingencies of individual development can bring out phenomena of evolutionary regression, where the wild animal that sleeps inside of each of us and has been muzzled by the processes of hominization, is likely to wake up and reveal our profoundly hybrid nature, angelical and beastly at the same time, and which corresponds to the image of the werewolf. This regressive potential has direct significance for representations of puberty as an event likely to allow the reemergence of anterior stages in the history of humanity, a pre-history of the subject as pre-human.
>
> (2004, 156)

Liminal positions and animal/natural correspondences

The age range studied in this chapter is a period of transition, which, as is often the case, not only brings about a call to order and a reformulation of gender identification but also constitutes a more general reminder of the ontogenic processes that shape a person. Several anthropologists have pointed out that there is a process of dehumanization and re-humanization that operates in many rites of passage (Van Gennep 1909; Turner 1969; Remotti 2002). Others have analyzed the power of plant and animal metaphors to signify the course of life, to mark gender and age, and to accompany transformations of the body by categorizing them according to functions that are perceived as "natural" (Verdier 1979; Teboul 2021).

Symbolic anthropology has long explored how liminal periods are threatened by the wild and are therefore socially assisted to establish a restrained and domesticated relationship with the wild. In *Les Pierres aux oiseaux*, Tina Jolas (1986) describes rituals like jumping from the rock on Saint Joseph's day – 19 March – or jumping over the St. John's bonfire on 24 June, or traditional practices in the Creuse region in France where "throughout adolescence a close and genuinely initiatory relationship is established between boys and birds. [This relationship is] expressed through a series of gestures: finding the nests, stealing the eggs, catching the nestlings" (Jolas 1986, 2). The idea is to learn the right use of one's senses and the pleasures involved in the activity. These senses and pleasures help the boy to become aware of the different times that define his life: the hunt symbolizes the beginning of adolescence, while the end of the hunt means entering the time of young adulthood.

As they hunt birds, the boys also learn how to handle trees, they carve out flutes, they "peg," "sew," "lift" and "pierce" the chestnut bark (ibidem). And in the same way, when they are adults, they reproduce these gestures when they find "wild stock" in the woods, those young trees that evolve without having been cultivated, but that grow thanks to the birds who disperse the pits of their fruit[3]: "A movement which, from one to the next, is inversed, thus drawing a cycle: birds pillage the orchards and men pillage the forest. To illustrate our claims, we should remember the mediating role of birds" (Jolas 1986, 7). In the rural European societies explored by Daniel Fabre, a boy needs to cross three thresholds in order to become a man:

> the first separates the domestic world (which is not only the world of home but also the social world) from the world of the wild; the second lies between the world of the living and that of the dead, and the third [which towers over the other two] between the masculine and the feminine.
>
> (Fabre 2019, 68)

These transitions happen through a metaphorical relationship with animals, specifically with birds, and that initiate young men in gender identities, in sexuality and love (Fabre 1986a, 2019). In the eastern Pyrenees, bird hunting and, later on, wild boar hunting, also contribute to the formation of a virile identity for young men. Daniel Fabre (1986b) magnificently describes the stages of this education in a place called Galamus, where eagle hunting and eating bacon, snails and strong-tasting meats grilled on a wood fire out in the forest are means to "absorb" the wild. This relationship with the wild reveals the aims of these specific social

3 Incidentally, it is interesting to note that in French, the word *sauvageon* (which is translated here as "wild stock" and also means "wild thing" or "little savage") is applied both to uncultivated saplings and to wild or uneducated children, and is moreover used in political discourse to stigmatize or exclude certain groups of youngsters with an immigration background, as did Jean-Pierre Chevènement, then Ministry of the Interior in 1999.

activities, which are linked to both nature and food, and shape masculinity, as they imply affirming oneself as a man, and using this awareness for one's amorous conquests. Learning masculinity, experiencing the end of childhood and the transition into young adult age are socially constructed in the French countryside through a very specific mastery of the territory and of the highly meaningful products of the wilderness (Fabre 1986b, 2005). Attending evening balls, gathering wild berries and maintaining that ancient relationship with the mountain hermit, who was considered the guardian of the margins that give access to life in the wild, are some of these practices.

More recently, in an enquiry about the parallel lives which were studied by Foucault, Philippe Artières tells us how "as a child" he had himself been "a savage in the Vosges" (Artières 2019, 26) and how, at the end of his teenage years, he had felt the need to take notes about this character called Jean. "In the eyes of the child that I was, Jean constituted a form of strong marginality that was neither linked to sexuality nor illegal practices, but rather to a certain relationship he had with nature," for instance, the fact that he "used to always bring along a piece of game meat" (Artières 2019, 29). The fascination this man holds for the young boys is linked to a secret connection he has with the animal world:

> we attributed almost supernatural powers to him, because the fact that he managed to catch those animals just seemed impossible to us. There seemed to be some mysterious relationship between him and the animals, because whenever he entered the pen, the boar would stop moving and would wait until he had left again to continue their wild stampede.
>
> (Artières 2019, 146)

These first-person narratives match our own research, which revealed the importance of forest cabin building in the east of France during the transition from childhood to adolescence.

This relationship with the wild has specific correspondences in urban contemporary societies. In popular fiction works found in the media, the metamorphosis of puberty is described through references to animal regression and a return to the wild (Moisseef 2010). In her work on critical age and the governing of liminal times in contemporary France, Ingrid Voléry analyzes how "when there is a break in time (because of an involution, a regression or an acceleration in the life course) we may expect [a] return to the wild" (Voléry 2016, 269). Bestiality and animalization, for instance through the figure of the wolf, are then called in to evoke a state of transition, a *persona liminalis* from which young men must distance themselves with the help of adults to go through a biographical transition (Moisseef 2010).[4] The image of the young wolf as a liminal person is also sometimes brought to bear by educators or other pedagogical staff to designate young people who have gone

4 The figure of the wolf also plays a role in women's age passages, from puberty to menopause, as Yvonne Verdier's work on tales (1980) has also shown.

off the rails. The image has also been assigned to forms of ethnicization that supposedly need to be "corrected" through cultural input. Thus, growing up is interpreted as a process of changing one's nature, with the assumption that the subject is exiting the condition of *homo selvaticus*.

Far fewer studies have looked at the opposite side of these age transitions, focusing on the end of adult age and exploring this relationship with animality. However, there are some elements in ancient medical literature. These writings were typically produced by doctors and authors who presented ageing as a specific process of "perfecting the moral and the spiritual character of a man" (Guyetant 1837, 39) by opposing this civilized subject to men who remain in the "animal" stage. In 1923, when the first sexual hormone extractions were carried out, Arthur Conan Doyle wrote a short story called *The Adventure of the Creeping Man* in which a 61-year-old emeritus professor falls in love with the daughter of one of his colleagues, with "the frenetic passion of a young man" (Conan Doyle 1923, 7). In order to find the appropriate sexual vigor to match his young wife's expectations, the professor injects himself with an invigorating serum that turns him into an ape. Conan Doyle's story is not just a tale of fiction: in the 20s, the French-Russian surgeon Serge Voronoff, who inspired this novel, transplanted tissues from monkeys' testicles into men, and then later monkeys' ovaries in women. Dalgalarrondo and Hauray retrace this medical experience, and reveal how anti-ageing therapies began with tissue transplants from animals to humans. They remind us how Brown-Séquard, one of the founders of endocrinology, in 1889 injected himself with extracts from guinea pigs' and dogs' testicles. At 72, he claimed that the injections had made him 30 years younger from the point of view of physical strength, intellectual capacity and sexual potency.[5]

The link between diseases that affect people in their fifties, sexuality and nature metaphors can also be found in Véronique Moulinié's *La chirurgie des âges* (1998). She documents masculine diseases of this age period, which were identified in South-Western France at the end of the twentieth century and refer to the loss of libido. The gout, a disease linked to the prostate or hemorrhoids, is dealt with in connection with the consumption of food that triggers sexual desire and quickens the bloodstream. The afflictions linked to "a very mobile virility [that] leaves its seat only to appear elsewhere" (Moulinié 1998, 249) are described as parts of a natural instability which seems to govern the masculine physiological rhythms during the adult age. This instability, which is seen as "natural," can be corrected and canalized: the image of the vine that needs to be pruned to stop the sap from rising is called forth and allows men to counter "the demon of midday" (*ibidem*). The analogy between man and the vine stock also appears in other studies, such as in the book *Jours de vigne* by Christiane Amiel, Giordana Charuty, and Claudine Fabre-Vassas (1981). The movements involved in pruning the vine translate the struggle of seduction that is fought out between different age groups, and which

5 We would like to thank Sébastien Dalgalarrondo and Boris Hauray who pointed out these scientific experiments to us.

especially opposes young and old men in a farandole that is held in Carcasonne each year on Saint Martin's day.

Passing an age thus implies interfering with a nature whose excesses must be contained. Going through a wild stage also constitutes a founding act in the construction of a person, hence the double inscription both in an age category and in a gender, and it is therefore a major foundation of the person's constitution. Indeed, the men we interviewed during our fieldwork permanently referred to animal metaphors to identify their masculinity and to talk about the food choices they made in order to accompany the changes their bodies undergo after the age of 40.

Animal metaphors and constructions of masculinity

These tropes continue to be active in today's cultures, though they have metamorphosed to correspond to wider social, economic and political transformations (Pizza 2013). While apparently far removed from folklore practices and preoccupations concerning the abuse of hormones,[6] and from symbolic systems that frame age transitions and the gender identification processes that go with them, animal metaphors and natural correspondences permeate the accounts of our interviewees. The men we met used these metaphors along three main strands. The first leads to an identification of the interviewees' food practices during adulthood with animality, specifically through domestic animals. The second line is mostly characterized in the interviewees' narratives by a desire to distance themselves, as adults, from the wild part of man through the embodiment of specific models of hegemonic masculinity. And the third line is rather linked to a return to the wild, which is connected to a removal of man from the domestic space and from "feminine" tasks and attitudes.

Moving toward the animal

The interviewees expressed the feeling of identifying with an animal as a sentient and intelligent being: "I ate pork, ok, but I have to recognize that when you look a pig in the eye, it's like a man!" exclaims Massimo at the end of the interview. Another vegetarian, Pietro, "refuses to feed [himself] with anything that has a heart." Stephan, who is the son of a mason father and of a mother who was a textile worker, remembers "the pig we used to sacrifice. . . . I can still hear the pig's piercing cries." The heart, the eyes and the cries all humanize the animal and take the interviewee back to the shared condition of living beings. Stephan still differentiates and opposes domestic or small-scale breeding, which is a part of "natural" life, to the bloody spectacle of industrial meat processing in which he worked when he was 17. For him, quitting that job after only a few months corresponds

6 This preoccupation takes on new shapes today, especially when it comes to estrogen and other hormones found in food destined for livestock.

to a more general life choice: working for the National Forestry Service, reading Ghandi, and taking an interest in macrobiotics. Later, when he started to work as a bricklayer he would be ridiculed by his teammates for his "chicken food," because at lunch he preferred to "eat vegetables."

Pigs, chickens and rabbits: the animals which are evoked here are mainly domestic animals, and through this domesticity, a different model of masculinity emerges. In this way, Stephan's food choices and behavior resonate with the way he chooses to live his relationship with his partner and the masculinity he wants to embody:

> *I had decided . . . not to do things like my father, I was in that protest movement where you'd spin your own wool, I'm the one who made the cardigan I wore at my wedding, and I was also part of those men who knew what it means to iron your own shirts. When we had children with my wife I was the one who changed the diapers. It seemed normal to me to get up at night, and to embark on a dynamic of sharing domestic tasks: my father, he'd come home from work, drink his aperitif, put his feet under the table and my mother had to prepare everything.*

This is an approach where the refusal of a profile of hegemonic masculinity meets a refusal of domination. Indeed, the fact of rejecting a dominant masculinity sometimes takes the shape of a caring form of masculinity which refuses patriarchal domination and the values that are usually associated with it, like strength, courage, independence or the ability to master one's feelings. These men identify their masculinity with other qualities that are a part of care, like sensitivity, interdependence and a relational attitude (Elliot 2016), which they develop during adult age. Tommaso says that he has been a vegetarian for five years (but he eats fish), and he is the father of two children aged 5 and 10. He pays a lot of attention to what his daughters eat: the brand Nutella is banished, as are other industrial snacks, sodas and chips, even for birthdays. His wife ironically calls him a fanatic when it comes to healthy food during adulthood, and she herself continues to eat meat at home. In spite of this, Tommaso is thrilled when he talks about the impact of his food education on his daughters, and relates how much they love vegetables, and omelet with leeks, which constitutes – in his view – an example for other parents who "are very happy when their children eat at home." In this context, it is once again a reference to an animal, in this case, an herbivore, that reveals another way of being a man:

> *You don't need meat to be strong and to be a man. The dinosaurs didn't eat meat. You know the story about the battle in which the warriors would fast before the battle to be stronger? Meat doesn't make you stronger. Me, after I've eaten meat, I just want to take a nap.*

Through this reference to domesticity, the animal metaphor allows for another identification with an alternative masculine model, which glorifies self-control and the ability to recognize the limits of the body. Furthermore, by evoking dinosaurs

Tommaso implicitly refers to a Paleo man and to a correspondence with wildness and to the promises of a return to a natural state (Dalgalarrondo and Fournier 2020) that existed before the advent of history. Hegemonic features like strength and body control are revisited to make room for other forms of masculinity which aim at optimizing the body during adulthood (Diasio and Fidolini 2019).

Rejecting the wild

A second cluster of tropes which describe our interviewees' relationship with animality refers more to a removal away from the wild, and which is also supposed to be a part of masculinity. The opposition between man and animal is here played out through dichotomies like wild-domesticated or predator-prey. Eugenio affirms

> I'm not a boar, which eats everything, I don't eat like an animal or make noises like an animal when I eat . . . you can't compare the man-boar who is from the countryside, working-class, who is a hunter, with someone like me: I am a professional, I am a doctor!

A class-based masculinity which is also urban and northern (Eugenio uses the stereotypes of obese families both from the South of Italy and from the countryside as counter-models) is thus differentiated from the *cinghialotti* (the small boars) that laugh at the fancy one and wonder about its obsession with staying in shape.

This distinction can, however, shift the predatory attitude linked to food consumption, and apply it to sexual behavior. In this way, Eugenio claims

> I don't show off the fact that I'm a man, a wolf, an alpha male by eating meat like other men, because I know that meat is not good for my health at this age: I do it by showing my girlfriend, who is twenty years younger than me, to the others!

The youth and the "freshness" of his masculinity are measured against the age of the sexual partner and helps Eugenio to certify that his food choices are good. This tragically ancient trope can also be formulated in terms of food: sexual "consumption" replaces meat consumption in order to affirm, otherwise, one's wolf status. While earlier Eugenio seemed to refuse to embody a normative model of dominant masculinity, he is here putting forward a masculinity that brings together many characteristics of a hegemonic profile: his virile heterosexual power, his predatory attitude toward women and his need of complicit masculine figures to confirm his dominant role in the homosocial dynamics (Fidolini 2020).

The reference to specific food habits as a tool to embody an urban and civilized masculinity also characterizes the account of Arduino. For him, optimizing his food mainly implies avoiding excesses. This wisdom makes him "take pleasure in consuming little portions of food and seeing [how] much [he can] count on his body." With the help of his wealthy social status and the cultural capital he has at his disposal, Arduino describes his current relationship to food as a continuous "research," which he defines as "intimate and spiritual" and which is shared with his wife, who is teacher as well. Thus, the relationship with food is interpreted as a search for spirituality and takes a central part in constructing an alternative model of masculinity based on body control:

> *When I eat with my friends, some of them [all are men] try to find out if I've already eaten the ham we bought together from our butcher, or they will eat quickly to see how much time it will take me to finish my plate. But it's a game, it's ironic . . . because I am more reflective than them . . . they finish every plate in two minutes and want to prove their voracity. . . . After all they are friends, they also have their kindness, when they are alone, they also eat a cream of mushroom soup, or water their flowers.*

His "civilized" behavior is exalted in contrast with the practices of his friends, who are indirectly linked to a bestial attitude toward food which Arduino rejects, in order to differentiate both his diet and his masculinity from those of his friends within homosocial commensality contexts.

Returning to the wild

The third figure we found is that of the man who returns to the wild as a form of transgression. When discussing the diet he eats with family at home, Giorgio describes this in terms of reduced quantities. On the contrary – and here he differs from Arduino – when Giorgio eats with his friends in homosocial situations, for example, during vacations or away from home, he displays a voracious attitude. In these circumstances, without his wife or children or mother,[7] Giorgio finds an opportunity to engage with "massive eating sessions":

> *I went on holiday with some friends, without our wives, and we ate like it was the last day of our lives. We ate just about anything. The lady from the hotel said she had never seen anything like it . . . we were like nine savage beasts.*

Just as for the young men described by Daniel Fabre (2019), homosociality and being outside of the domestic space are part of the construction of a virility (Flood 2008) that goes through a form of returning to the wild.

However, the image of the man who goes back to the wild also appears in another form, again based on taking one's distance from the domestic space and the feminine sphere. During fasting, Vincent says he "de-socializes a bit" from commensality. This attitude is highly criticized by his partner. Nevertheless, fasting seems to have a positive influence on the way his masculinity is shaped, with useful effects on the relationship with his partner. Surpassing his limits, controlling his

7 Like many of the Tuscan men we interviewed, Giorgio has lunch at his mother's house or drops by to pick up dishes that she has prepared.

body and having a particular contact with nature represent specific qualities that Vincent confers to his fasting practice, and which also reinforce his identification with a hegemonic model of masculinity. According to Vincent:

> *there are women who are attracted to this. With my partner, we have spoken about my fasting and what she finds attractive about me and my habits. She mentioned this side of me that is a bit wild, the man who needs to leave her on her own from time to time. She said this fascinates her. I think that the fact that I need to be on my own, that I fast, that I need to be alone to write or to read, is part of my virilization.*

Through "virilizing" fasting, Vincent is able to place himself in a dominant position in the gender hierarchy. His masculinity is defined by his capacity to control his body and to push himself to overcome its limits. This practice not only allows the recovery of a kind of original "purity" and "naturalness," as the interviewee says. It can also confirm, in a different way, a profile of hegemonic masculinity which challenges social conventions and is able to return to the wild in order to better affirm its powerful virility. This return to the wild is superposed on the image of the man who is revitalized by being alone in nature. The autarchy of the body that feeds itself with its own dregs reflects the affirmation of an acting singularity, which does not need the assistance of other people during its mature age. The passage through illness, this temporary walk along the ridge that separates life from death, can sometimes result in a return to nature that is perceived as a potential for salvation or healing.

Conclusions

These narratives show that the food habits described by the men we interviewed are often assimilated into gender identifications processes that uphold specific representations of a masculine connection with nature and wildness. Nonetheless, at the heart of the three narratives on masculinity which we analyzed, the relationship with the wild is not only expressed in terms of a departure from the wild before the return to culture, as initiation processes often seem to describe (Turner 1969). It actually produces a plurality of identifications that are expressed in age transitions, and which men try to control. Beyond the frequently discussed association between the consumption of red meat and virility (Hell 1988), the forms of masculinity which we found in our fieldwork bring to bear images of "fellowship" (Vialles 2005, 85) that establish a dialogue between herbivores and carnivores, vegetarians and omnivores, and beyond that, the vegetable and the animal worlds. These oppositions give rise to different forms of masculinity that reveal hybrid models of gender (Demetriou 2001): indeed, a dominant masculinity does not present itself as a unique and fixed block but rather as an open and changing space where hegemonic characteristics can coexist with weakness, concern and lack.

Masculine performances are played out based on a dialectic of opposition between self-regulation and the absence of limits, with the animal trope being invoked either to represent an incorporated discipline that enhances physical strength, virility, willpower and self-control, or as a deterrent against a hegemonic and predatory masculinity. This polar opposition thus presents a view of nature that could be called tame, and that differs from a nature where the wild deploys itself uncontrollably.

In the same way as for youth (Fabre 2019), in the passage between adult age and old age, the border between the domestic and the wild is doubled by the various ways in which we think about the masculine and the feminine. The tropes of the wild allow to voice the relationship with the feminine world. We identified three images, each representing the topic of wildness in a different mode: the domesticated animal, the predator and the lonesome hero who returns to the wild.[8]

In the first narrative mode, which we mainly analyzed through the case of Stephan who identifies with the "sacrificed pig," the relation to women is tackled by establishing a closer connection between masculinity and domesticity, resulting in a process of hybridization of the masculine, through the identification of the masculine with domestic animals. In the second strand of narratives, women are rather seen as the object of a masculinity that becomes predatory and identifies with the metaphor of the wolf in order to retrieve its lost virility, or, at the least, a virility that is called into question. In the third narrative mode, women seem excluded both through references to the homosocial world and by the importance given to man's solitude and self-sufficiency, which present a "different" construction of strong and dominant masculinity based on the metaphor of the solitary hero who returns to nature and to the wild.

While it may seem that the years between 40 and 60 represent a period in which men's attitudes to food and their bodies assume a different meaning from that of their earlier years, masculine performances take up polyvalent signifiers through natural and animal metaphors. The narratives of the interviewees confirm that individual and collective constructions of masculinity are precarious processes that need continuous reconfirmation. In the accounts of the men we met, this reconfirmation seems to be found in their different – and sometimes contradictory – rhetorical connections with the wild and the wildness. During their midlife transition, the interviewees seem to view their masculinity as being called into question, and therefore seek new patterns to identify with to pursue alternative hegemonic profiles. Animal metaphors are thus both called forth and rejected to protect men from the passing of time and from bodily changes.

8 We do not have enough elements at our disposal to put forward any interpretation, but it would be interesting to compare the figure of the ageing man to Fabre's sketches of young men who do not completely succeed their age transition and remain "stuck" beyond one of the borders that had to be crossed: they either stay on the side of the non-domestic and become "men of the wild" or they continue to mix with supernatural beings (the "cookies" or the "crazies" as Fine [2021] puts it), or they remain on the feminine side (the "effeminate"). These models hybridize and mingle, but they have a heuristic value in the fact that they allow us to think transitions that are not stabilized in our society.

Appendix: Tables of the interviewees

Table 9.1 List of the interviewees recruited in France (Grand Est)

Name	*Age*	*Social profile*	*City*	*Type of diet*
Antoine	42	Shop assistant	Strasbourg	Vegetarian
Claude	440	Mechanical technician	Mulhouse	He does not declare any specific diet
Christophe	558	Journalist	Mulhouse	Vegetarian
Damien	440	Professor (university)	Strasburgo	He does not declare any specific diet
Éric	446	Farmer	Small village of the Alsatian countryside	He does not declare any specific diet
Luca	447	Restaurant owner	Strasbourg	He does not declare any specific diet
Hamid	445	Retailer	Strasbourg	He does not declare any specific diet
Maurice	551	Insurant agent	Strasbourg	He does not declare any specific diet
Pierre	557	Music teacher	Strasbourg	Vegetarian
Quentin	559	Freelance retailer	Mulhouse	Vegetarian
Stephan	559	Teacher (high school)	Strasbourg	Flexitarian
Thomas	558	Factory worker	Strasbourg	He does not declare any specific eating regime
Valentin	442	Factory worker	Metz	Vegan
Vincent	663	Software engineer	Strasbourg	He does not declare any specific diet

Table 9.2 List of the interviewees recruited in Italy (Tuscany)

Name	*Age*	*Social profile*	*City*	*Type of diet*
Arduino	55	Teacher (high school)	Florence	Flexitarian
Diego	57	Office worker	Florence	He does not declare any specific diet
Eugenio	50	Dentist	Small town near Florence	Vegetarian
Flavio	42	Small business entrepreneur	Small town near Arezzo	Fruitarian
Giorgio	43	Factory worker	Florence	He does not declare any specific diet
Ivo	52	Security guard	Livorno	Vegetarian
Leonardo	51	Engineer	Florence	He does not declare any specific diet
Lino	55	Restaurant owner	Small town near Florence	He does not declare any specific diet
Massimo	61	Teacher (high school)	Small town near Florence	He does not declare any specific diet
Mirko	51	Factory worker	Arezzo	Vegan

Name	*Age*	*Social profile*	*City*	*Type of diet*
Pietro	58	Factory worker	Small town near Florence	Vegetarian
Pino	58	Factory worker	Small town near Arezzo	Flexitarian
Tommaso	45	Hairstylist	Small town near Arezzo	Vegetarian
Tancredi	56	Small business entrepreneur	Florence	He does not declare any specific diet
Vincenzo	54	Office worker	Florence	He does not declare any specific diet

References

Adams, Carol. 1990. *The Sexual Politics of Meat: A Feminist-Vegetarian Critical Theory*. London: Continuum.

Amiel, Christiane, Girodana Charuty, and Claudine Fabre-Vassas. 1981. *Jours de vigne. Les femmes des pays viticoles racontent le travail*. Villelongue d'Aude: Atelier du Gué.

Artières, Philippe. 2019. *Le dossier sauvage*. Paris: Verticales.

Bertrand, Julien, Martine Court, Christine Mennesson, and Vinciane Zabban. 2015. "Introduction. Socialisations masculines, de l'enfance à l'âge adulte." *Terrains & Travaux* 27(2): 5–19.

Bessin, Marc, and Marianne Blidon. 2011. "Vieillir." *Genre, Sexualité et Société* 6, https://journals.openedition.org/gss/1979.

Blatterer, Harry. 2007. "Contemporary Adulthood: Reconceptualizing an Uncontested Category." *Current Sociology* 55(6): 771–792.

Charlap, Cécile. 2019. *La fabrique de la ménopause*. Paris: CNRS.

Conan Doyle, Arthur. 1923. *L'homme qui grimpait*. www.jeremy-brett.fr/cariboost_files/conan_doyle_homme_qui_grimpait_im.pdf.

Connell, Raewyn. 1995. *Masculinities*. Cambridge: Polity Press.

Dalgalarrondo, Sébastien, and Tristan Fournier. 2019. "Introduction. Optimization as Morality, or the Routes of the Self." *Ethnologie Française* 176(4): 639–651.

Dalgalarrondo, Sébastien, and Tristan Fournier. 2020. *L'utopie sauvage. Enquête sur notre irrépressible besoin de nature*. Paris: Les Arènes.

Delanoë, Daniel. 2006. *Sexe, croyances et ménopause*. Paris: Hachette.

Demetriou, Demetrakis Z. 2001. "Connell's Concept of Hegemonic Masculinity: A Critique." *Theory and Society* 30: 337–361.

Depecker, Thomas, Anne Lhuissier, and Aurélie Maurice, eds. 2013. *La juste mesure. Une sociologie historique des normes alimentaires*. Rennes: Presses Universitaires de Rennes.

Diasio, Nicoletta. 2004. "Le plaisir de se gouverner: les grignotages enfantins entre biopouvoir et techniques de soi." In *Le corps. Le sensible et le sens*. Edited by Gilbert Vincent, 47–62. Strasbourg: Presses Universitaires de Strasbourg.

Diasio, Nicoletta, and Vulca Fidolini. 2019. "Garder le cap. Corps, masculinité et pratiques alimentaires à l'âge critique." *Ethnologie Française* 176(4): 751–767.

Diasio, Nicoletta, and Virginie Vinel. 2007. *Il tempo incerto. Antropologia della menopausa*. Milan: Franco Angeli.

Digard, Jean-Pierre. 1993. "Les nouveaux animaux dénaturés." *Études Rurales 129–130*: 169–178.

Douglas, Mary. 1970. *Purity and Danger. An Analysis of Concepts of Pollution and Taboo*. Harmondsworth: Penguin Books.

Elliot, Karla. 2016. "Caring Masculinities: Theorizing an Emerging Concept." *Men and Masculinities 19*(3): 240–259.

Fabre, Daniel. 1986a. "La voie des oiseaux. Sur quelques récits d'apprentissage." *L'Homme 26*(99): 7–40.

Fabre, Daniel. 1986b. "Le sauvage en personne." *Terrain* 6 http://journals.openedition.org/terrain/2891.

Fabre, Daniel. 2005. "Limites non frontières du Sauvage." *L'Homme 175–176*: 427–443.

Fabre, Daniel. 2019. *L'invisible initiation*. Paris: Éditions de l'EHESS.

Featherstone, Mike, and Mike Hepworth. 1985. "The Male Menopause: Lifestyle and Sexuality." *Maturitas 7*: 235–246.

Fidolini, Vulca. 2020. "Heteromasculinities. Sexual Experiences and Transition to Adulthood among Young Moroccan Men in Europe." *Men and Masculinities 23*(2): 242–265.

Fidolini, Vulca. 2022. "Eating Like a Man. Food, Masculinities and Self-care Behavior." *Food, Culture & Society 25*(2): 254–267.

Fine, Agnès. 2021. "Invisible Initiation." In *Daniel Fabre, l'arpenteur des écarts. Actes du Colloque de Toulouse février 2017*. Edited by Nicolas Adell, Agnès Fine, and Claudine Vassas, 111–127. Paris: Éditions de la Maison des Sciences de l'Homme.

Fischler, Claude, and Estelle Masson, eds. 2008. *Manger. Français, Européens et Américains face à l'alimentation*. Paris: Odile Jacob.

Flood, Michael. 2008. "Men, Sex and Homosociality: How Bonds between Men Shape Their Sexual Relations with Women." *Men and Masculinities 10*(3): 339–359.

Foucault, Michel. 1988. "Technologies of the Self." In *Technologies of the Self. A Seminar with Michel Foucault*. Edited by Luther H. Martin, Huck Gutman, and Patrick Hutton, 16–49. Amherst: The University of Massachusetts Press.

Foucault, Michel. 1989. *Résumé des cours 1970–1982. Conférences, essais et leçons du Collège de France*. Paris: Julliard.

Foucault, Michel. 1997. *Il faut défendre la société. Cours au Collège de France 1976*. Paris: Gallimard and Seuil: Éditions de l'EHESS.

Gough, Brendan. 2007. "Real Men Don't Diet: An Analysis of Contemporary Newspaper Representations of Men, Food and Health." *Social Science and Medicine 64*(2): 326–337.

Gullette, Margaret. 1998. "Midlife Discourses in Twentieth-century United States: An Essay on the Sexuality, Ideology and Politics of Middle-ageism." In *Welcome to Middle-age (and other Cultural Fictions)*. Edited by Richard A. Shweder, 3–44. Chicago: Cambridge University Press.

Guyetant, Sébastien. 1837. *Le médecin de l'âge de retour et de la vieillesse, ou conseils aux personnes des deux sexes qui ont passé l'âge des quarante-cinq ans*. Paris: Ballard.

Hearn, Jeff. 1995. "Imaging the Aging of Men." In *Images of Aging. Cultural Representations of Later Life*. Edited by Mike Featherstone and Andrew Wernick, 97–114. New York-London: Routledge.

Hell, Bertrand. 1988. "Le sauvage consommé. Classification animale et ordonnance cynégétique dans la France de l'Est." *Terrain 10*: 74–85.

Hepworth, Mike, and Mike Featherstone. 1998. "The Male Menopause. Lay Accounts and the Cultural Reconstruction of Midlife." In *The Body in Everyday Life*. Edited by Sarah Nettleton and Jonathan Watson, 277–301. New York-London, Routledge.

Jolas, Tina. 1986. "Les pierres aux oiseaux." *Terrain* 6 http://journals.openedition.org/terrain/2892.
Julier, Alice, and Laura Lindenfeld. 2005. "Mapping Men onto the Menu: Masculinities and Food." *Food & Foodways 13*: 1–16.
Lalive D'Epinai, Christian, and Stephano Cavalli. 2007. "Changements et tournants dans la seconde moitié de la vie." *Gérontologie et Société 121*: 45–59.
Lax, Jacob B., and Angela G. Mertig. 2020. "The Perceived Masculinity of Meat: Development and Testing of a Measure across Social Class and Gender." *Food, Culture & Society 23*: 416–426.
Méchin, Colette. 1985. "Essai d'anthropologie de l'alimentation dans le nord-est français." *Revue des Sciences Sociales de la France de l'Est 14*: 79–95. www.persee.fr/doc/revss_0336-1578_1985_num_14_1_2836
Meloni, Paolo. 2011. *Modi giusti. Cultura materiale e pratiche di consumo nella provincia toscana contemporanea*. Pise: Pacini Editore.
Moisseeff, Marika. 2004. "Le loup-garou ou la virtualité régressive du pubertaire masculin." *Adolescence 22*(1): 155–171.
Moisseeff, Marika. 2010. "Apprivoiser la métamorphose pubertaire." *Ethnologie Française XL*(1): 75–84.
Moulinié, Véronique. 1998. *La Chirurgie des âges. Corps, sexualité et représentations du sang*. Paris: MSH.
Mycek, Mari Kate. 2018. "Meatless Meals and Masculinity: How Veg* Men Explain Their Plant-based Diets." *Food and Foodways 26*(3): 223–245.
Nath, Jemal. 2011. "Gendered Fare? A Qualitative Investigation of Alternative Food and Masculinities." *Journal of Sociology 43*(3): 261–278.
Parasecoli, Fabio. 2005. "Feeding Hard Bodies: Food and Masculinities in Men's Fitness Magazines." *Food and Foodways 13*: 17–37.
Pizza, Giovanni. 2013. "Gramsci e Di Martino. Appunti per una riflessione." *Quaderni di teoria sociale 13*: 75–120.
Remotti, Francesco. 2002. *Forme di umanità*. Milano: Mondadori.
Roos, Gun, Ritva Prättälä, and Katriina Koski. 2001. "Men, Masculinity and Food: Interviews with Finnish Carpenters and Engineer. *Appetite 37*: 47–56.
Schilling, Chris. 1993. *The Body and the Social Theory*. London: Sage.
Schwaiger, Elisabeth. 2009. "Performing Youth: Ageing, Ambiguity and Bodily Integrity." *Social Identities 15*(2): 273–284.
Sobal, Jeffery. 2005. "Men, Meat, and Marriage: Models of Masculinity." *Food and Foodways 13*(1–2): 134–158.
Sumpter, Kristen C. 2015. "Masculinity and Meat Consumption: An Analysis through the Theoretical Lens of Hegemonic Masculinity and Alternative Masculinity Theories." *Sociology Compass* 9: 104–114.
Teboul Jeanne. 2021. "Du 'pioupiou au rapace': ambivalences initiatiques de la formation au métier militaire." In *Daniel Fabre, l'arpenteur des écarts. Actes du Colloque de Toulouse février 2017*. Edited by Nicolas Adell, Agnès Fine, and Claudine Vassas, 173–181. Paris: Éditions de la Maison des Sciences de l'Homme.
Turner, Victor W. 1969. *The Ritual Process. Structure and Anti-Structure*. London: Routledge.
Vainionpää, Kirsi. 2005. "Male Menopause." The Birth of a New Illness? *Acta Electronica Lapponiensis 109*: 75–88.
Van Gennep, Arnold. ([1909] 1960). *The Rites of Passage*. London: Routledge.

Verdier, Yvonne. 1980. "Grands-mères si vous saviez . . . Le *Petit Chaperon Rouge* dans la tradition orale." http://expositions.bnf.fr/contes/cles/verdier.htm

Verdier, Yvonne. 1979. *Façons de dire, façons de faire. La laveuse, la couturière, la cuisinière*. Paris: Gallimard.

Vialles, Noëlie. 2005. "La viande ou la bête." *Terrain 10*: 86–96.

Vinel, Virginie. 2007. "Andropause et ménopause: une comparaison des images de la vulgarisation médicale." In *Féminin, masculin: anthropologie des catégories et des pratiques médicales*. Edited by Virginie Vinel, 199–222. Strasbourg: Cahiers du Portique.

Voléry, Ingrid. 2016. *Âges critiques et temps liminaux dans la France contemporaine. Contribution à une sociologie du cours de vie* (Habilitation à diriger des recherches). Lille: Université de Lille 3.

10 "Heavy food" and "being in nature"

Revisiting male manual workers' narratives on work, food and health

Gun Roos

Nicoletta Diasio and Vulca Fidolini examine in their chapter the trope of wildness and how animal metaphors shape masculine food practices and midlife transitions among French and Italian men. They conclude that animal metaphors are both drawn on and rejected to protect men from ageing and bodily changes. In this commentary, I will return to some earlier qualitative studies on men, occupation and health lifestyles in Finland and Norway, which I participated in around the turn of the twenty-first century (Roos, Prättälä, and Koski 2001; Prättälä 1997; Roos and Wandel 2005; Wandel and Roos 2005). I will, on the basis of these studies, explore how sociocultural factors are reflected in how men talk about food practices, health and their bodies. The studies in Finland and Norway had a parallel design and all included semi-structured interviews with male manual workers (carpenters and loggers), aged 40 to 50 years. I will here describe the roles of work, health and the connection to nature in the narratives of these men from two Nordic countries.

Manual work requires heavy food

The capacity to work has a central role in Nordic welfare states; men have traditionally been breadwinners in the families and work has been important for men's identity. The interviews with manual workers in Finland and Norway focused on thoughts and practices related to health and started with an occupational biography and questions related to working conditions. In addition, most of the interviews were done at the workplace. Thus, it is not surprising that work and working conditions had a central role in the narratives.

The men who worked at construction sites expressed that they above all value their work, because it gives them a sense of autonomy, they experience that the work is fairly varied and they appreciate the concrete results. They conveyed that they can at least, to some degree, plan their own work schedule and that being a carpenter gives them freedom to move around and work on different construction sites. Working outdoors was also mentioned as an asset and important reason for their choice of occupation. However, there were as well accounts of showing resentment with the hierarchy at construction sites and lack of control. They were also aware of that the hard workload may lead to damage and many could feel the body wear and tear, but they didn't want to complain and show weakness.

DOI: 10.4324/9781003243496-19

Autonomy, self-sufficiency and a feeling of being in control of your work and body seemed important. A recent Australian study of men's eating habits based on occupation indicates that autonomy and control still are important (Carroll, Capel, and Gallegos 2019).

The Finnish and Norwegian manual workers seemed to often take eating for granted; they said that they mainly eat out of necessity to get energy for work and that they need to work to get food on the table. As a Norwegian carpenter said: "*I'm not interested. I just want a lot and eat it fast so that I can go out and do something else. I think it's the same case for many men in general.*" One of the Finnish carpenters described as part of his occupational biography why he in his youth moved to Sweden: "*I got out of the army and had no home to go to. I didn't want to stay there to be fed. I had to go somewhere and they were shouting after men in Sweden . . . there was a big need to get additional bread.*" Whereas it was voiced that if you do not work you need less, as an unemployed Finnish carpenter said: "*you can manage on less because you don't consume at work.*"

In line with Pierre Bourdieu's (1984) theories claiming that the working classes emphasize body strength and see the body in a more instrumental way, many of the Finnish and Norwegian manual workers acknowledged the masculine metaphor of the body as a machine needing fuel. As a Norwegian carpenter said: "*Think about a car that needs both gasoline and oil, and if it doesn't get that, it will stop functioning. It is a tremendous wear on the body if it doesn't get the right nourishment.*" The men expressed that the body needs the right fuel for manual labor and that working outdoors in the cold weather also demands something heavier than, for example, vegetables and soup. As a Finnish carpenter said: "*It has to be something stronger, it has to do with the occupation. The muscles burn energy a lot more than those who have lighter work.*" The men often supported the view that, to do heavy manual work, the food also has to be heavy and the portions have to be big enough to give a feeling of being full. However, it seems that some men felt that they had to defend their views and that the general dietary recommendations to eat more vegetables and less fat were not in line with their energy needs. Some joked about the consequences for work performance if they would only eat salad. As a Finnish carpenter said:

> *you need some more energy than if you for example eat just vegetables you start to feel weak in a week (laughter) but when it's in general always is that you should eat a lot of vegetables and avoid everything with even a little fat but I need fat.*

Construction sites rarely have canteens for workers, and most brought their own food or bought something in nearby shops. They expressed that they were not comfortable using regular cafés or canteens, because the portions were too small and expensive for manual workers, but also because of their often dirty working clothes. As a Finnish carpenter said:

> *it depends if there are canteens with good cheap food nearby, when you are in overalls you don't feel like going to all canteens, and the food in all canteens is not meant for manual workers, portions are so small that it's not worth to pay the price.*

Some men also showed resentment with that their bosses had lunch coupons and eat lunch in cafees: "*the builders have those free coupons, lunch coupons, they go in the café to eat but workers do not go, at least I have not seen anyone go anywhere to eat.*"

Ambivalence toward healthy eating

The manual workers in Finland did not, when we talked with them about food in the 1990s, seem to feel that they needed to defend meat consumption, but they recognized that eating is closely related to health. When they were asked about a healthy diet, they often emphasized variety, modesty and needing energy for work: "*people need because they work hard and they need what they use. I eat almost anything, but regularity and required amount and variety are all relevant factors.*"

The Nordic countries have, for many decades, focused on the relation between nutrition and health and prevention of diet-related diseases. They have been nutrition-policy forerunners, and the first formal nutrition policies were adopted already in the 1970s and 1980s (Roos, Lean, and Anderson 2002). In Finland, as a result of the growing emphasis on risk factors and prevention of cardiovascular diseases, the North Karelia project was launched in 1972 as a pilot community intervention project to reduce mortality from coronary heart disease in Finland (Puska et al. 1995). Many of the interviewed Finnish carpenters came from rural areas, including North Karelia, where diets traditionally were high in dairy fats and low in vegetables. The influence of tradition was evident in the men's opinions about high fat foods being necessary for heavy physical labor. Work and particularly the physical demands of manual work were often brought up as resistance or a legitimate justification for not fulfilling the general norm of eating healthy. For example, some felt that they didn't have time to eat breakfast in the morning before work, and if they had to work late, dinner was postponed. The men proposed that manual workers can allow themselves to eat snacks and unhealthy foods, because the calories are used up, and they cannot work effectively if they are hungry. But they also recognized that the traditional Finnish diet they grew up on was not very varied: "*it was fairly one-sided food in the 1950s it was pork in Eastern Finland and fish of course when we were at the cabin by the lake.*"

Some of the men were interested in healthy lifestyles and had a positive attitude toward health recommendations. Some had got advice to change their diets because of health problems such as high cholesterol levels or high blood pressure. Although traditional working-class men are in general considered not to be interested in written instructions about healthy life, the men's attitudes toward health were heterogenous. Ritva Prättälä (1997) concluded in her study of Finnish loggers and carpenters that they were "between a rock and a hard place." The men could be said to be between opposing forces; they described that tradition, working conditions and family were main elements that affected their health, and that each of these elements could both promote and prevent healthy lifestyles. Therefore, the group was heterogeneous in their orientations toward healthy lifestyles. The men described how carpenters have traditionally considered that heavy manual work requires heavy food, and thus, some of them expressed defiance toward dietary guidance and joked about how they could not work if they would follow the nutritional recommendations. But this ambivalence was not expressed by all; some of the men were interested in healthy lifestyles and had changed their diets. The family often had a positive influence on health; the men said that they worked to provide for their family and described how their wives had informed them about healthy eating and cooked them healthy meals including more vegetables and less fat.

Nature provides

Nicoletta Diasio and Vulca Fidolini explore in their contribution how men used animal metaphors along three strands. They describe that one of the narrative strands gives importance to man's solitude and self-sufficiency and masculinity based on the metaphor of "the solitary hero who returns to nature and to the wild." This metaphor was also used by the men in the Nordic countries. For example, the study in Finland included interviews with carpenters with rural backgrounds working in the Helsinki area. Many had grown up in poor families that lived of small farms and forestry, and during their holidays, they were at their rural summer cottages or visited their relatives in the countryside. At the time when they were young in the 1970s, there was a rapid urbanization and young people left the countryside to find jobs in the cities or abroad. A Finnish carpenter described that his first work experience was in forestry, that he in particular appreciated the autonomy and that he could decide his working hours and work pace. He further told that it gave him a feeling of freedom and his description reflected how much he appreciated the beauty of being close to nature:

> *it was working as a logger. You could decide your working hours and work pace . . . it was good with the freedom and you could be in nature all the time. And a morning during springtime, like this, it was really nice to start the morning in the forest and the birds were always singing when you went to work and you had time to listen before you started the saw.*

Some of the Finnish carpenters said that they gather and eat "wild" food; they described that they have wild game meat in the freezer; they said that they liked fishing and that they pick wild berries and mushrooms in the forest. A carpenter portrayed how they have had wild berries for an evening snack: "*I put cloudberries to thaw and we ate before we went to bed, we picked them in the summer.*" Some among those who didn't fish or hunt themselves talked about relatives who had provided them with wild moose meat to stock in the freezer and potatoes to store in the cellar. Fishing was popular among some of the men. Many had grown up in the lake district and were used to both fishing and eating fish. One of the Finnish carpenters described how he thinks that being in nature and fishing is fundamental for a healthy lifestyle:

> *"Once a summer to visit Kuusamo and go fishing it's for me the best (laughter) To swirl the salmon fly. It's the goal that I've been charging myself for eleven months. If you catch one fish or have one bite then its's a hell of a good one."*

Conclusions

Revisiting 30-year-old narratives on food practices by male manual workers in two Nordic countries show that the participating men aged 40 to 50 years were especially inspired by work and connection to nature when talking about food. They described that they work to get bread on the table and that, to be able to work, they need the right type of energy or fuel for the body. Self-sufficiency, autonomy, freedom and variety were given importance. The narratives suggest that occupation and working conditions may shape and limit men's food practices. The men seemed to position themselves as laid back, passive and traditional, and when talking about food, work and daily routines were more important than health concerns. The Finnish carpenters that we interviewed in the 1990s were born close to nature, they appreciated that working outdoors gave them continued connection to nature, and in their leisure time, they returned to the wild free nature. Nature for them provided food, peace of mind, health and feelings of freedom.

What about today? Would we expect to get similar narratives if we would interview male workers aged 40 to 60 years in Nordic countries today? Most likely work would still have a central role when the men talk about food, but because many of the men would have grown up in urban areas, they would probably not have had the same continuous connection to nature. The role of health and climate for food practices would likely also be given some more emphasis among workers. They would most likely talk more about cooking, because surveys in Nordic countries show that men cook more and especially working-class men have started cooking more since the turn of the century (Holm et al. 2015). The men would, as the men in the 1990s, still be between opposing forces both promoting and preventing healthy lifestyles, and their accounts would draw on demands related to the work situation and the sociocultural background, level of knowledge and education.

References

Bourdieu, Pierre. 1984. *Distinction: A Social Critique of the Judgement of Taste*. London: Routledge & Kegan Paul.

Carroll, Julie-Anne, Eleanor M. Capel, and Danielle Gallegos. 2019. "Meat, Masculinity, and Health for the 'Typical Aussie Bloke': A Social Constructionist Analysis of Class, Gender, and Consumption." *American Journal of Men's Health* November–December: 1–12.

Holm, Lotte, Marianne Pipping Ekström, Sara Hach, and Thomas Bøker Lund. 2015. "Who Is Cooking Dinner? Changes in the Gendering of Cooking From 1997 to 2012 in Four Nordic Countries." *Food, Culture & Society* 18: 589–610.

Prättälä, Ritva. 1997. *Puun ja kuoren välissä. Metsurit ja kirvesmiehet puhuvat terveelllisestä elintavoista (Between a Rock and a Hard Place. Loggers and Carpenters Talking About Healthy Lifestyles)*. Helsinki: Publication of the LEL Employment Pension Fund, 32.

Puska, Pekka, Jaakko Tuomilehto, Aulikki Nissinen, and Erkki Varitainen. 1995. *The North Karelia Project: 20 Year Results and Experiences*. Helsinki: The National Public Health Institute (KTL).

Roos, Gun, Michael Lean, and Annie S. Anderson. 2002. "Dietary Interventions in Finland, Norway and Sweden: Nutrition Policies and Strategies." *Journal of Human Nutrition and Dietetics 15*: 99–110.
Roos, Gun, Ritva Prättälä, and Katriina Koski. 2001. "Men, Masculinity and Food: Interviews with Finnish Carpenters and Engineers." *Appetite 37*: 47–56.
Roos, Gun, and Margareta Wandel. 2005. "I Eat Because I'm Hungry, Because It's Good, and to Become Full": Everyday Eating Voiced by Male Carpenters, Drivers and Engineers in Contemporary Oslo. *Food and Foodways 13*: 169–180.
Wandel, Margareta, and Gun Roos. 2005. "Work, Food and Physical Activity. A Qualitative Study of Coping Strategies Among Men in Three Occupations." *Appetite 44*: 93–102.

Conclusion

The wild in gastronomy and beyond[1]

Sébastien Dalgalarrondo and Tristan Fournier

On a rewilding laboratory

"Would you like some more live ants with your yogurt?" We are in a restaurant, but not in Southeast Asia or Amazonia where eating insects is common. We are in Denmark, in Copenhagen, and this is a phrase that those who have been lucky enough to eat at the highly prestigious Noma restaurant may have heard. René Redzepi, this restaurant's atypical chef, offers the perfect incarnation of the renewal of the wild in global gastronomy. By including yogurt flavored with live ants on one of his menus, this unusual chef invites us to discover the taste potential of this local ant species collected by hand and delicately scattered on a homemade yogurt. The insects, judiciously trapped in this creamy dish, are just waiting to be chewed to release their formic acid, allowing diners who have become their predators for the day to enjoy a delicious lemony flavor. Food critics, struggling in the face of so many innovations and deviations, have difficulty describing his approach. American food writer David Rosengarten has suggested calling this cuisine "primitivist modern." This oxymoron captures the disruptive quality of this chef, who is often presented as one of the most influential in the world.[2]

The culinary experience at Noma goes far beyond the gastronomic art: the aim is to offer diners a real anthropological experience, which requires them to overcome their disgust, question their culinary preconceptions and reposition themselves in the long and complex history of the human diet. The recipes using ants on the Nordic Food Lab's[3] online blog perfectly exemplify this will for a radical revival of our appetencies:

> On a whittled liquorice root we brushed honey infused with juniper wood and studded it with the frozen ants, freeze-dried fruits, seeds, toasted crushed grains, small aromatic leaves and flowers. This "chimp stick," as it came to be

1 Translated by Cadenza Academic Translations.

2 In 2021, Noma was awarded its third star in the Michelin Guide. For the fifth time, it was named "World's Best Restaurant" by The World's 50 Best Restaurants.

3 A culinary think tank arising from the collaboration between gastronomic entrepreneur Claus Meyer and Noma's head chef René Redzepi.

DOI: 10.4324/9781003243496-20

> known, was served on a charred piece of juniper wood . . . and guests were able to choose whether to suck all the edible parts off the stick, mixing the sweet, sour, aromatic, bitter and toasted notes or to pick at it and taste the flavours individually. Our hope was to evoke the mess of flavours and textures that would surround the ants in their field or forest habitat, and to use the sour taste to brighten the palate for the ensuing menu.[4]

The use of this "chimp stick" is not insignificant. Eating by sucking on the stick allows the diner to reconnect with their own animality by inviting them to temporarily abandon their anthropocentric position as a superior outsider: "inspired by ant and termite 'fishing' tools used by chimpanzees. Eating insects like these is part of our evolutionary heritage."[5]

As conscientious sociologists and curious eaters, we obviously would have liked this conclusion to include an ethno-aesthetic analysis of this restaurant, its menu, and the measures devised by this unconventional chef to encourage the "average eater" to make this primitivist leap. However, to get a table at such a renowned restaurant, you have to book well in advance, which is not always the easiest thing to manage when carrying out field research, especially with the added complication of Covid-19 restrictions!

To carry out our exploratory investigation in this Danish laboratory, we chose to take another route, far from the gastronomic market and its select clientele. We chose to look at the MAD Foundation (whose name means "food" in Danish), also created by Redzepi himself, which is like a highly original tool for democratizing his approach. At the core of this ambition for democratization is a phone app called VILD MAD ("wild food" in English), which allows its users to discover good sites for collecting wild food: plants, berries, mushrooms, crustaceans, etc. We can see that the primitivist approach is closely entwined here with digital modernity. However, as stated by our interlocutor, the one who is in charge of developing this app and whom we interviewed in Copenhagen (who is himself a very experienced forager), this is not about survivalism. The aim is not to make everyone capable of living totally autonomously in the woods, but rather to help laypeople rediscover their immediate environment and learn to forage. It is to give everyone the opportunity to reconsider their everyday diet by reconnecting with the wild and its discreet abundance:

> Denmark is not a land of gastronomy like France. We don't have a highly developed culinary culture, but that doesn't mean we have no assets. Take our apples, for example. They need longer to ripen, they are slower-growing, and

4 "Ants and a Chimp Stick," Nordic Food Lab, June 2, 2013, accessed March 16, 2022, http://nordicfoodlab.org/blog/2013/6/ants-and-a-chimp-stick/.

5 "Ants and a Chimp Stick," Nordic Food Lab, June 2, 2013, accessed March 16, 2022, http://nordicfoodlab.org/blog/2013/6/ants-and-a-chimp-stick/.

> these specific climate conditions give our apples a very distinctive taste. The challenge is to change how we look at our products.

One very interesting detail emphasized by our passionate interlocutor is that the application never tells users exactly where to find a particular plant or wild berry, since the aim is to encourage novice foragers to wander, to explore their natural environment by themselves and thereby to create their own culinary topography. This is the opposite of the supermarket model. The aim of such an app is not to generate wild food consumers 2.0 but to use digital technology and its appeal (particularly to younger generations) to help laypeople take ownership of their immediate environment and explore it according to their tastes, the seasons, their preferred routes and their family culinary history. Redzepi is perfectly aware of the limitations of a nostalgic discourse on children:

> The old-fashioned charms of the outdoors are competing with the newfangled baubles of tech. Kids are so comfortable with devices that the ability to marry it with analog tech – your two feet – and propel you out the door and into the wild is a nice conflation of old and new.[6]

The objective is to allow these children and their parents, by capillary action, to reconnect with the primitive pleasure of seeking food, of foraging in nature, like a hunter-gatherer who, day after day, constructs their own mental map of their territory and identifies the resources available according to the seasons, the plants that are present and the weather conditions.

Also offering many training courses to chefs from around the world, the MAD Foundation has developed a highly original program to introduce schoolchildren to wild cuisine. This is how R. Redzepi presents it:

> The basic premise is simple: everyone in the world should grow up as a forager. . . . Knowing your ABCs in nature, the flora and fauna, the patterns of the landscape and the rhythm of the seasons is as important as learning how to read and write. How might the world look different, if we all were foragers? We hope that our program can be an inspiration for others to do the same in different countries and cultures. If they see how much we depend upon it, and if they grow up loving it then they will fight to take care of it.[7]

6 Redzepi, quoted in: Jason Tesauro, "One of the World's Most Famous Chefs Wants to Help You Forage for Your Food," *Washington Post*, June 29, 2017, accessed March 16, 2022, www.washington-post.com/lifestyle/food/one-of-the-worlds-most-famous-chefs-wants-to-help-you-forage-for-your-food/2017/06/29/b59a1d34–5ceb-11e7–9fc6-c7ef4bc58d13_story.html.

7 Redzepi, quoted in: Jason Tesauro, "Noma Chef Rene Redzepi Wants to Help You Forage with His New Vild Mad Food App," Good Food, July 11, 2017, accessed March 16, 2022, www.goodfood.com.au/recipes/news/noma-chef-rene-redzepi-wants-to-help-you-forage-with-his-new-vild-mad-food-app-20170711-gx8r1z.

By analyzing this educational program and through practical experience of the MAD Foundation's teaching, we were able to understand this democratization of the wild, which operates beyond the purely gastronomic sphere. The foundation has trained several dozen "VILD MAD rangers" in Denmark, so that they can provide training year-round for school groups. One of these rangers was kind enough to welcome us, so that we could discover the content of this training in "real conditions." Our investigation took us near to the sea, to a poor suburb 40 minutes from Copenhagen, where we tried out this introduction to wild food. Leaving the station, we noticed that we were in a highly urban environment with lots of residential buildings, a commuter suburb very different from the splendor of the capital. It didn't feel like a good place for foraging, and the presence of a KFC and numerous fast-food restaurants all around the station showed us how far there was to go. . . . However, after walking for just a few minutes, the landscape changed radically. We entered an unspoiled area, where there were few cars. We saw dunes in the distance and we noticed that a lot of birds seemed to be enjoying the tranquility of this space. Now, we just needed to find the building where our training would take place. We walked across a zone specializing in sailboat repair, and even though the idea of this experience filled us with enthusiasm, the surrounding vegetation did not seem conducive to foraging. This feeling was exacerbated by the fact that it was December. It was cold, fine rain made us fear the worst, and it wasn't really the ideal season, but the Covid-19 epidemic had forced us to adapt our research. According to our GPS, we were just tens of meters from the meeting place. We saw a modest modern building, its external walls covered with green metallic panels. It was nothing like the forest hut we had imagined the day before when planning our journey. We hurried along the little path, with no signs to reassure us. We were in danger of being late and there was nobody around. We heard the sound of waves, the sea was close, and as we walked around the building looking for the entrance, we came face to face with a fox, which ran away when it saw us. For sociologists searching for the wild, this was a reassuring sign, and we began smiling again! Through the window, we saw the ranger who had agreed to teach us. He was putting wood in a stove and he invited us in. On his chest, he wore a "VILD MAD" badge and as we began talking, we soon realized that he was intrigued by our presence. He didn't really understand why we were there and confirmed to us that it wasn't really the right season. We began to doubt the relevance of our approach. Our ranger, whose name was Olaf,[8] invited us into the room where we would be taught, and as we were looking around, we immediately noticed the incongruity of the place. We glanced at each other knowingly, the look of sociologists who know they are on the right track. We had no doubt that the Danish laboratory of wild food was there, in front of us.

We were expecting to find a professional kitchen, with the associated layout, furniture and hygiene standards. In fact, the space was like a living room with a convivial atmosphere, its walls decorated with a collection of stuffed birds and

8 The ranger's name has been changed to ensure anonymity.

small rodents, as if we were in a natural history museum. In the room were small island units where participants could work in groups. On one of these, the equipment for cooking was laid out. We also noticed that there were aquariums. On the right, Olaf showed us an educational resource designed to familiarize pupils with mussel farming techniques. At the back, we saw a stainless-steel sink and a dishwasher. The place was difficult to describe, and we can only say that it was profoundly hybrid. As you can imagine, Olaf was not wearing a chef's hat! He was the manager of this Nature Center and a trained forestry and landscape engineer, recently promoted to the rank of MAD ranger. We spent a few minutes introducing ourselves to each other, and we very quickly recognized our interlocutor's teaching skills. Aged around 50, he was remarkably modest, and wore a broad smile at all times, particularly when we encountered translation issues (the interview was conducted in English, and all of us had our limitations in the language). Olaf portrayed his training in wild cuisine with the MAD Foundation as a major transformation! Even though he knew his area and its natural resources perfectly, he had never imagined the flavor potential and cooking possibilities they offered:

> In my family, we made a few wild berry preserves, but it was a declining tradition, mainly restricted to older people. This training changed many things in my personal life. I discovered lots of edible plants and their unbelievably diverse tastes, textures, and colors. I also learned to pay attention to the presentation of dishes. Before, I ate out of bowls, and it was often the same thing: pork, potatoes, cabbage, etc. This training taught me that the presentation of dishes was very important, and that I needed to pay attention to the appearance as well as the taste.

He had adopted the first rule of any good chef: we eat with our eyes first.

Our time on the beach was short: the weather was gray and windy, there had been a freeze the night before and there were still patches of snow in places. There wasn't much to gather. We started by collecting seaweed, and Olaf encouraged us to look at it carefully to really become familiar with it:

> With school groups, we focus on just three or four plants. Our aim is not to see everything, but rather for the children to leave with the ability to forage for three or four plants with their friends or family with total confidence!

He taught us to taste "like rabbits," putting the product between our teeth and top lip, to give us a chance to really get to know and absorb the specific taste of the seaweed. Then, we went to some bushes that still had a few berries on them: we filled our wicker baskets with rose hips (*Rosa canina*) and hawthorn berries (*Crataegus monogyna*). The last product we collected was a small, bright yellow berry: sea buckthorn (*Hippophae rhamnoides*). These abundant fruits cheered up the gray, wintry atmosphere. "It's the end of the season," he told us, but these berries actually needed to be harvested at the start of winter, at the time of the first frosts. We learned that sea buckthorn berries ripen before winter, but that they are almost impossible to harvest at this stage, because they are too fragile: they burst between the fingers as soon as you try to grasp them. Frost therefore becomes the forager's unexpected ally: they can cut branches full of berries (allowing the shrub to regenerate) and harvest them easily by gently tapping the branches into a container so that the berries fall off. Suddenly, coming right in the middle of winter made sense! This original botanical story in which frost becomes the forager's assistant was like a real moment of enchantment for us. We saw these somber bushes covered in yellow berries in a new light and we could easily imagine the effect such a discovery might have on the receptive brains of young children. On our return to the "laboratory" and after carefully washing our harvest, we got ready to cook. We noticed two incongruous products on the table that our ranger had prepared ready for us to make our "meal": industrial mayonnaise and crackers for spreading. These items were far from wild, but we quickly understood that this original assemblage allowed children to make the leap, to take "the risk" of eating an unknown wild product, to enculture it, so to speak. These everyday food products allow them to "domesticate" their harvest, to integrate it into reassuring eating routines. Once again, Olaf emphasized the need for the food to be visually appealing and well presented on the plate before we tasted it. The green seaweed on the crackers stood out attractively against the yellow, red and purple berries. We were hungry and thoroughly enjoyed the food, which was original to say the least. The juice that we made with the rest of the berries was delicious, and its flavor, color and acidity contrasted with the preceding culinary experience.

We had now experienced this original laboratory of wild food for ourselves. The trainer was a hybrid being – half forest ranger and half chef. We had experimented in a familiar setting, very far from an impressive professional kitchen, and the wood-burning stove at the back created a feeling of being at home. The introductory course was highly practical and pragmatic. What mattered was doing things for oneself, creating emotions and new sensations while still maintaining a strong connection with the familiar. Before we said goodbye to Olaf, he offered us the opportunity to talk to some children who had experienced the same course. We met them at their school. They were 12 years old and the teacher kindly allowed us half an hour to talk to them outside, despite the cold (this was in the middle of the Covid-19 epidemic). When we asked the children about their parents' interest in this experience, several of them said it had revived memories and family traditions. Some of them were immigrants from the Mediterranean region, and Olaf told us on the way back that the subject of foraging had allowed him to start a dialogue with these foreign mothers who had foraged as children. In this respect, the laboratory takes on a new dimension: foraging becomes a means of intercultural exchange and reciprocal enrichment. Most of the children had downloaded the VILD MAD app on their cell phones and some of them used it with their families during Sunday walks. These children above all emphasized the fact that this food was free and focused on the possibility of survival: "It's useful if you get lost in the forest." Thus, they gave us a feeling of having acquired a new skill, which is particularly reassuring since it offers a possible route to autonomy. A few days later, when we were dining in one of the restaurants in Copenhagen that boasts this wild approach (the menu is imposed), to our great surprise and satisfaction we found a bright yellow mix in our dessert that tasted like our recent discovery: sea buckthorns! The wild had found the most perfect way of following us. Indeed, this social laboratory of wild food had taken the form of an interstitial and disruptive (ruptural) "real utopia" (Wright 2013).

The wild in contemporary food: assemblage and incarnation

The rewilding of food can be explored in terms of a promise, a nourishing utopia that allows us to enrich our everyday eating while preserving the nutritional diversity required for the survival of our species, through its practice and the reflexive act that this consumption implies. Wild food sits at the intersection of two normative injunctions that are decisive for contemporary eating: improving one's health and protecting the environment:

> Diversity is our starting point and our goal. It forms a loop of feedback mediated by ecology, necessity, and appetite. There is no single food that can nourish us on its own. The pursuit of good food is in itself also the pursuit of biocultural diversity, the pursuit of a future where everyone can not only eat but eat well. Diversity – of autochthonous genetic material, of organisms, of cultural practices, of ideas – is what keeps our world at its most resilient and robust. It is the cloth that weaves our fates together and upon which we share our meals.[9]

By adopting the analytical perspective of the sociology of promises (Fournier and Lepiller 2019; Dalgalarrondo and Hauray 2015; Pollock and Williams 2010), we want to understand new zones of exchange (Deleuze and Guattari 1980) and the social mechanisms that can initiate adhesion to the promise of the wild. The description of the "Danish laboratory" has allowed us to highlight the assembled dimension of this promise: our initiation took place in a hybrid space, an original arrangement of symbols and representations, midway between a natural history museum (stuffed animals) and a domestic kitchen (sink, island units, crackers, and a jar of mayonnaise), not forgetting the contribution of digital technology to improving people's connection with the wild. The second mechanism at work in the performativity of this promise is incarnation: in addition to the educational processes used, our trainer was the perfect incarnation of the possibility of transforming oneself and one's everyday life through contact with wild food. He described his training with the MAD Foundation to us as a turning point in his life: a rewilding process marking a profound transformation of the self and of domestic commensality, because he was also a father who had taught his children. Let us look at these two mechanisms in more detail.

9 "Who We Are," Nordic Food Lab, accessed March 16, 2022, https://nordicfoodlab.org/whoweare/.

Assemblage

First of all, we can observe that these promises take a specific form: the individual is faced with an original assemblage (Ong and Collier 2008; Delanda 2016) of discourses, images, facts, fictions, imaginaries and practices. Following the works of Deleuze and Guattari (1980), it appears that these assemblages (which must be seen as a process, an arrangement in the making) function like a tiled roof, each element overlapping with those around it. At its heart is a cluster of fundamentals, argumentative registers that can capture attention and allow everyone, depending on their preconceptions and interests and their capacities and skills, to explore the promise's potential. A child who is reluctant to cook can cling to the familiar presence of mayonnaise on the table; another might find that the physical and sensory dimension of foraging outdoors helps them to connect with the experience. It is through this symphony of words and things that everyone can write their own score. The analysis of the promise of the wild and the question of its adoption in everyday life highlights the effects of this assembled dimension. From this perspective, the contemporary practice of "wellness fasting," as examined in conversation three between S. Dalgalarrondo, T. Fournier and O. Bidet, provides an archetypal example. This voluntary abstinence corresponds to the possibility of a "transhumanism with bare hands": a biotechnological honing through the exploration of a "primitive" technique of the body. The hope of the biology of aging is combined here with a desire to reconnect with our "true nature," with the optimal physiology of the Paleolithic hunter-gatherer, and/or (depending on one's leaning) with our wild companions (Haraway 2003). An exploration of the self and of our capacities for resistance that is akin to a political act, a momentary liberation from the feeling of hunger with the aim of rediscovering the ways of a frugal, simple and minimalist life. This assembled form of the promise helps to spark, justify and claim a self-rewilding process. Conversation one by O. Lepiller, O. Roudelle, S. Dury and S. Dalla Bernardina on the contemporary consumption of game meat also demonstrates this aggregate dimension. The promise borne by this wild meat gives those who eat it a multitude of ways of justifying their atypical practices. There is of course the health argument, based on the nutritional properties of large game meat, but the authors show that this works in symbiosis with the evocation of heritage protection and preservation, of a food culture that allows them to claim a rootedness and an unfailing connection to a territory and the creatures that live in it. All of these arguments allow them to justify eating game, in a context that is a priori unfavorable to it, in which people are urged to become flexitarian, and in which vegetarianism and veganism are on the rise. The complex assemblage analyzed in conversation two between C. Pineau and L. Mariani on natural wine allows us to expand the question of circulation within the promise and of the inclusive potential of this cognitive wandering. Natural wine offers those who drink it many anchor points: science and its tools, such as microscopes, combine with more esoteric or even spiritual practices, reflected perfectly by the concept of "biodynamism." Here, once again, we have an assemblage that can appeal to various audiences and highly diverse preferences. Instrument-based rationality (e.g., monitoring fermentation processes) forms an original pairing with the quest for sensitivity and the possibility of a demanding and respectful dialogue with the wild dimension of vines. These multiple entry points into the promise of natural wine offer many possible routes for those concerned to explore, but above all, they make the drinker an actor in this assemblage in their own right. They cannot remain at a distance, like a spectator watching a symphonic concert might choose to focus on the sound of a particular instrument. Those who drink natural wine, through the act of tasting it, their expectations about the wild nature of this wine and their proximity to the unstable and fluid (de Laet and Mol 2000) product they are consuming, are called upon within the assemblage as fully fledged actors in its construction.

Incarnation

Redzepi offers an incarnation of wild food, particularly when he chooses to pose proudly on social networks holding a dead wild duck (mallard) in his hand, exhibited like a trophy. Many of Redzepi's enthusiasts were astounded by the controversy surrounding his Game and Forest Menu, which in fall 2019 included a mallard-based dish in which all the parts of the animal were used (including the brain as tempura in the head, the heart tartare in the beak, and the leg presented with its feathers intact). Certain critics, offended by so much realism, even called it a "trash dish." The sous-chef at Noma, Pietro Leemann, gave a particularly interesting response, because it showed that this aesthetic of rewilding, far from being a Tarantino-like "trash" aberration, is actually assumed and claimed as a political act, a meaningful incarnation:

> The reason why this dish was made is simple. When we eat meat (we use very little at Noma!), there is always a death in the middle, whether you are required to see it or not. To celebrate this death, we decided to pay respect to the animal by using everything, from head to legs just to not waste anything. However much splatter this may be in the eyes of many, I assure you that Mallard is less stained with blood than the chicken breast wrapped in the boxes of plastic that you buy at the supermarket, not to mention the low-priced and off-season vegetables that you always buy in the supermarket at home.[10]

The process of "incarnation" (Hauray and Dalgalarrondo 2019) is thus to be understood as a modality of discourse, in this case, an approach to food as the foundation for an approach to life. More precisely, it is a mode of veridiction: incarnation is a mode of truth-telling that can be distinguished from other discourses presenting things as true. We hypothesize that in a situation of uncertainty, when

10 "Noma replies after chef says Redzepi went 'too far' with duck dish," Fine Dining Lovers, January 22, 2020, accessed March 16, 2022, www.finedininglovers.com/article/noma-duck-head-dish.

many people are trying to identify the right way to eat, this process of incarnation can have a significant influence on people's willingness to experiment. These experimentations will then produce new knowledge and trigger new "incarnations." Incarnation can thus play a crucial role in the performativity of promise discourses. We consider the term "incarnation" to be more appropriate than "embodiment," because:

> incarnation refers more exclusively to a living body (incarnation comes from the Latin word *incarnare*: in- "caro," to make into flesh or to be made flesh), whereas embodiment has been used to analyze how expectations are "inscribed" into objects or machines (Borup et al. 2006). Moreover, its religious etymology is of interest here (in Christian theology, the doctrine of Incarnation holds that through Jesus, "the Word was made flesh"). . . . Four characteristics of "incarnation" can be analytically distinguished to explain its capacity to impact beliefs and behaviors: (a) it mobilizes individual stories, (b) it relies on individuals' own experimentation . . ., (c) it depicts ability to do things . . ., (d) it often involves displaying bodies.
>
> (Hauray and Dalgalarrondo 2019)

The body displayed can be that of the rewilded eater, or even that of the animal that will be eaten/incorporated. In this regard, the incarnation of the rewilding promise sheds light on "parrhesia." As underlined by Foucault (2011) in his analysis of the different forms of veridiction in Antiquity, the specificity of this mode of truth-telling is that the parrhesiastes speaks in their own name, about an actual and concrete situation, and that the effectiveness of their discourse relies on the fact that they stake their own life on what they are saying. The parrhesiastes is fundamentally someone who resists, who refuses to submit and uses the development of a personal ethics to question the world and find their place in it. For Foucault, submission and parrhesia are totally incompatible. Through their practice, the parrhesiastes is engaged in a power struggle. By developing a technology of the self, based here on the development of a particular diet, they try to achieve freedom from servitude, to found the ethics that allows them to become master of themselves. For the subject, the aim is to break free from norms, in order to identify a still unlikely but desirable way of being, via this reflexive work. Finally, incarnation produces knowledge of a specific epistemic value: "It allows a connection between the two dimensions of life identified by the French philosopher Canguilhem (1966): *le vivant* (the living) and *le vécu* (the lived), two grammatical forms of the verb *vivre* (to live)" (Hauray and Dalgalarrondo 2019). The rewilded eater, by consuming a certain type of food, by involving their body in an uncertain or even risky practice, can make eating a political act. They become a thought-provoking model for others by providing the "living" and concrete proof of a new way of life. This brings us back to the concerns of men experiencing the midlife crisis analyzed by N. Diasio and V. Fidolini in conversation five. In response to this life event that transforms bodies and requires dietary changes, some of their survey subjects mobilize the registers of the animal and the wild to compensate for a diminished masculinity and reassert certain lost qualities: physical performance, autonomy, virility, etc. These same qualities were extolled by the manual laborers that G. Roos interviewed, who argued that they had to eat in a way that reflected their lifestyle and occupation, namely by consuming heavy food and being in nature. The analysis of urban foraging by F. Paddeu and F. Roussel in conversation four also sheds light on this process of incarnation. The practice of those who forage for reasons other than economic hardship or necessity, seems to correspond more closely to the demand for a way of life than to a pragmatic search for a new source of food supplies. These foragers fit into a nourishing utopia. They too are trying to rewild themselves in contact with nature. However, their participation in this practice seems to be primarily motivated by politico-aesthetic issues, by the desire to incarnate a different mode of existence, dictated in particular by environmental considerations. The aim is not to nourish themselves on a physiological level, but to nourish the "self" through contact with nature and with the knowledge required for this practice. The analysis of freegans as urban foragers by S. Mégarbané in her commentary seems to be consistent with this. The political dimension of this practice is omnipresent, particularly because it challenges the concept of the commons and because the bodily engagement – namely the incorporation of the products collected, which are potentially contaminated by urban activities – is like a form of manifesto. Urban foragers are an incarnation less of the possibility of "life in capitalist ruins," in the words of Tsing (2015), than of the urgent need for an urban model that maintains the possibility of personal contact with the wild. Freegans display a highly original side of the parrhesiastes seeking rewilding: these postmodern "gatherers," who play with the risk of food poisoning by salvaging discarded products from dumpsters, make incorporation a political act. Like those who fast, they occupy the dietary margins, in order to transform themselves (the concept of immunoevolution) and at the same time they give weight to their alert. Taking health risks strengthens and supports the political act.

Food (along with its promises) represents an area of analytical opportunity that is still underexploited by those interested in our changing relationship with nature and in the forms of discreet engagement that climate issues are creating in our kitchens. The objective of this book is to explore some of the possibilities of such a connection: a nature considered here at its closest to everyday life, as a potential way of rewilding the self, at the heart of infra-politics (Scott 1990), in the privacy of domestic and routine practices. A wild nature that, here, is no longer just an abstract concept to be deconstructed but the concrete substance of a future digestive process.

References

Borup, Mads, Nik Brown, Kornelia Konrad, and Harro Van Lente. 2006. "The Sociology of Expectations in Science and Technology." *Technology Analysis & Strategic Management 18*(3–4): 285–298.

Canguilhem, Georges. 1966. "Le concept et la vie." *Revue Philosophique De Louvain* 64(82): 193–223.

Dalgalarrondo, Sébastien, and Boris Hauray. 2015. "Les économies de la promesse anti-âge. Le cas de la DHEA." *Sciences Sociales et Santé* 33(2): 5–30.

De Laet, Marianne, and Annemarie Mol. 2000. "The Zimbabwe Bush Pump: Mechanics of a Fluid Technology." *Social Studies of Science* 30(2): 225–263.

Delanda, Manuel. 2016. *Assemblage Theory*. Edinburgh: Edinburgh University Press.

Deleuze, Gilles, and Félix Guattari. 1980. *Capitalisme et schizophrénie. Volume 2. Mille plateaux*. Paris: Éditions de Minuit.

Foucault, Michel. 2011. *The Courage of the Truth (The Government of Self and Others II): Lectures at the Collège de France, 1983–1984*. Translated by Graham Burchell. Edited by Frédéric Gros, François Ewald, and Alessandro Fontana. London: Palgrave Macmillan.

Fournier, Tristan, and Olivier Lepiller. 2019. "Se nourrir de promesses. Enjeux et critiques de l'introduction de deux innovations dans le domaine alimentaire: Test nutri-génétique et viande in vitro." *Socio* 12: 73–95.

Haraway, Donna J. 2003. *The Companion Species Manifesto: Dogs, People, and Significant Otherness*. Chicago, IL: Prickly Paradigm Press.

Hauray, Boris, and Sébastien Dalgalarrondo. 2019. "Incarnation and the Dynamics of Medical Promises: DHEA as a Fountain of Youth Hormone." *Health: An Interdisciplinary Journal for the Social Study of Health, Illness and Medicine* 23(6): 639–655.

Ong, Aihwa, and Stephen J. Collier, eds. 2008. *Global Assemblages: Technology, Politics, and Ethics as Anthropological Problems*. Malden, MA: John Wiley & Sons.

Pollock, Neil, and Robin Williams. 2010. "The Business of Expectations: How Promissory Organizations Shape Technology and Innovation." *Social Studies of Science* 40(4): 525–548.

Scott, James C. 1990. *Domination and the Arts of Resistance: Hidden Transcripts*. New Haven, CT and London: Yale University Press.

Tsing, Anna Lowenhaupt. 2015. *The Mushroom at the End of the World. On the Possibility of Life in Capitalist Ruins*. Princeton, NJ: Princeton University Press.

Wright, Erik Olin. 2013. "Real Utopias." *Politics & Society* 41(2): 167–169.

Index

Note: Page numbers in *italics* indicate a figure and page numbers in **bold** indicate a table on the corresponding page. Page numbers followed by "n" indicate a note.

For Product Safety Concerns and Information please contact our EU
representative GPSR@taylorandfrancis.com
Taylor & Francis Verlag GmbH, Kaufingerstraße 24, 80331 München, Germany

www.ingramcontent.com/pod-product-compliance
Ingram Content Group UK Ltd.
Pitfield, Milton Keynes, MK11 3LW, UK
UKHW021307260926
475702UK00048B/816

* 9 7 8 1 0 3 2 1 5 2 9 1 2 *